YUL
MTL

Published by Applied Research + Design Publishing
An Imprint of ORO Editions
Gordon Goff: Publisher

www.appliedresearchanddesign.com
info@appliedresearchanddesign.com

Copyright © University of Montreal 2015

All rights reserved. No part of this book may be reproduced, stored in a retrieval system, or transmitted in any form or by any means, including electronic, mechanical, photocopying of microfilming, recording, or otherwise (except that copying permitted by Sections 107 and 108 of the U.S. Copyright Law and except by reviewers for the public press) without written permission from the publisher.

This book is printed in four colors with spot varnish applied to all photographs on 157gsm OJI matte art paper. The ends are also printed in four colors on 160gsm woodfree paper.

You must not circulate this book in any other binding or cover and you must impose this same condition on any acquirer.

Book design by Circular Studio
Art direction and design: Pablo Mandel
Design: Maureen Höllboll
French text typesetting: Horace Pozzo

Production Assistance: Meghan Martin
Authors: Philippe Poullaouec-Gonidec, Sylvain Paquette, Patrick Marmen
Cover image: Gilles Hanicot, 2011
Back image: CPEUM/CUPEUM, 2014

10 9 8 7 6 5 4 3 2 1 First Edition

Library of Congress data available upon request. World rights available.

ISBN: 978-1-940743-09-7

Color Separations and Printing: ORO Group Ltd.
Printed in China.

International Distribution:
www.appliedresearchanddesign.com/page/distribution

YUL MTL
Moving Landscapes

Philippe Poullaouec-Gonidec
Sylvain Paquette
Patrick Marmen

ar+d
APPLIED RESEARCH
+DESIGN
PUBLISHING

Contents

Acknowledgements
/ 7

Preface
/ 8

Foreword
/ 10

Introduction
/ 14

1
Context
/ 34

Entering Montreal From A Highway Perspective

2
Vision
/ 68

Montreal's International Gateway Corridor

3
Illustration
/ 88

Ideation Process For A Regional Vision

4
Tools
/ 158

Principles, Criteria And Planning Scenarios

Conclusion
/ 244

Bibliography
/ 256

Authors
/ 260

YUL/MTL
Philippe Poullaouec-Gonidec, Sylvain Paquette and Patrick Marmen

This publication was made possible thanks to funding from
the ministère des Transports du Québec (Canada).

Acknowledgements

The authors wish to thank the people who contributed to the production of this book, namely: Valerie Gravel, Research Officer (CPEUM) for her contribution to the manuscript and the editing of the illustrations, as well as Anne-Sophie Frican (Research Assistant - CPEUM) and Julie Bergeron (Research Officer - CPEUM) for assisting in the research.

Philippe Poullaouec-Gonidec, Full Professor (Senior Researcher) and Sylvain Paquette, Associate Professor (Co-researcher) also wish to express appreciation:

to Louis-Philippe Roy, YUL/MTL Project Manager, as well as to all the members of the Monitoring Committee from the Direction de l'Île-de-Montréal du ministère des Transports du Québec (Canada), namely: Anne Pelletier, Assistant Director, Bureau de projet Turcot; Marie-Élaine Rochon, Information Officer; Sylvie Tanguay, Research Officer and Linda Jasmin, Head of Division;

to the Bureau du design de la Ville de Montréal for carrying out the WAT_UNESCO Montreal 2011 and to all members of the Autoroute 20 gateway corridor work table, namely: City of Dorval; City of Montreal (Bureau du plan; Division de l'urbanisme; Borough of Côte-des-Neiges - Notre-Dame-de-Grace); Borough of Lachine; Borough Le Sud-Ouest; Borough of Ville-Marie; City of Montreal West; City of Westmount; Conférence régionale des élus de Montréal; Communauté métropolitaine de Montréal; Ministère des Affaires municipales, des Régions et de l'Occupation du territoire; Ministère des Transports du Québec - Direction de l'île de Montréal; Ministère du Tourisme; Aéroport de Montréal; Agence métropolitaine de transport; Canadian Pacific; Canadian National; Parks Canada (Lachine Canal);

to the YUL/MTL project research team (CPEUM/CUPEUM) primarily funded by the ministère des Transports du Québec (Canada) as well as by the City of Montreal, namely: Julie Bergeron, Research Officer; Rachel Bonin, Administration Officer; Catherine Brouillette, Technical Support; Marc Chenouda, Research Officer; Lyndsay Daudier, Project Assistant; Valerie Gravel, Research Officer; Karina Luiza Desmarais, Administrative Assistant; Sophie Lacoste, Research Assistant; Patrick Marmen, Research Officer and Christine Robitaille, Research Assistant;

And to the two partners of the Chair in Landscape and Environmental Design, being the ministère des Transport du Québec (Direction de la recherche) and the Université de Montréal (Bureau Recherche, Développement et Valorisation, Direction des relations internationales, Faculté de l'aménagement).

Preface

In 2011, when I was in Canada, Philippe Poullaouec-Gonidec unexpectedly called me to replace an "excused" member of the jury at the international ideas competition, *"YUL/MTL: Moving Landscapes."*

The competition was "open in regards to its considerations", it imposed no "preconception", no method, no budget, no planning. "Everyone", without specialty or expertise, questioned the vast "disused" territory that lies between Montreal's international airport, YUL, and downtown Montreal.

Transport infrastructure and brownfield sites mainly took up the territory. The landscape was also full of neglected pieces, because of "anarchical" dispositions. With its central location, it provided considerable prospective potential, which could produce many new innovative urban expressions.

I did not know the site or the subject of the competition. After a quick, and therefore incomplete, visit of the 17-km-long territory in question, I put myself in the role of president of the jury.

What saved me was that the competition was international, therefore open to "foreigners" like me , who were unfamiliar with the habits, customs, and issues of the area. My eyes were imbued with freshness!

The subject was discovered.

It concerned the themes of "city entrances", the crossing of infrastructures whether inhabited or not, as well as their burying. It also concerned the memory, the resurgence of buried original geographies' fragments needing to be brought back to the surface. It concerned the recapture or intensification of neglected areas, a vast recurring theme for the 1960s infrastructure of large cities. It also concerned the conversion of projects. One in particular, the Montreal entrance named the "Turcot Interchange", was presented as a huge three-dimensional spaghetti dish; it had to be "demolished" because of fatigue from too many freezing and thawing cycles.

By scraping a little, we could also be concerned about managing time to implement temporary projects that are related to events, agriculture, culture or landscapes as we wait.

Themes are infinite: each competitor had to cross and live in these spaces, to feed the thoughts that were submitted to our jury.

The forward-looking utopias were also welcome – we, the jury, had the duty to potentiate them by placing them forward. Expectations do not have any operational purpose, but rather help redefine a new collaborative planning process with both public and private stakeholders present on this territory.

My experience in a jury about "issues" helped me animate the discussions and prevent the "operational" ones – always part of this kind of jury – from confiscating the subjects.

I set clear, shareable rules.

First, respect for the work of all participants, whoever they may be, by taking into consideration all elements submitted to us. This very tedious consideration process was not only to respect, but also to exploit, the "tank of ideas" to better enlighten our judgments and serve the Montrealers.

The members of the jury had to respect one another and listen because that was the rule.

We conducted this fascinating work stuck for three days in the Canadian Centre for Architecture (CCA), and we reached a unanimous consensus on the selection of winners and honorable mentions.

Our last duty was to establish and write our methodology and questions, and then justify our choices.

In this way, all future *YUL/MTL: Moving Landscapes* exhibitions will be prepared, the book included, so that the concepts and solutions we continue to create will be unlike anything we imagined.

Édouard François, int.FRIBA
RIBA & DPLG Architect, Urban Planner ENPC, a.e. ENSBA
International Fellow of the Royal Institute of British Architects
Chevalier de l'Ordre des Arts et des Lettres in France

Foreword

How do the results from an ideation process stemming from an international ideas competition and a design workshop create the backbone of a collaborative planning project? The present work intends to demonstrate that creative engagement in a territorial planning process assembles and brings together the territorial stakeholders' viewpoints while allowing visualization and implemention of that vision.

It is important to stress straightaway that this initiative of the Chair in Landscape and Environmental Design from the University of Montreal (CPEUM) and of the UNESCO Chair (CUPEUM) from the same institution was held with the support and collaboration of the ministère des Transports du Québec. In this respect, this department is a pioneer in Québec, and even in North America, supporting, in the context of a highway project, the key role of ideation through consultation, mobilization, and building awareness amongst territorial stakeholders. By focusing on inclusive ideation work that is not meant to retain a general idea but an "atlas of possibilities", as suggested by the jury of the *YUL/MTL: Moving Landscapes* ideas competition, directed by Édouard François, Architect and Urban Planner, the position of this approach will help to ensure a dialog between everyone.

An original use of this *"atlas of possibilities"* is put forward in this book to formalize the design principles and criteria, as well as to create sensible and coherent urban planning scenarios with a view toward implementing future projects along an infrastructure corridor.

This question of the meaning given to the territory is one of the essential challenges of a requalification project for the city entrance highway corridor. Therefore, the *YUL/MTL: Moving Landscapes* project reminds us of the importance of determining this requalification project on a territorial basis, because mobility carries in itself an experience that involves placing the crossed and lived places onto a landscape.

The initiative presented in this publication thus contributes to shedding new light on current thought about landscape, urban mobility, and the city entrance concept. We hope that the contribution and benefits from this collaborative planning exercise can inspire the implementation of future projects in Montreal and the development of similar processes elsewhere in the world.

Yet, this approach already fits into the objectives of the Montreal UNESCO City of Design designation. "Neither label nor consecration, (this nomination) is an invitation to develop Montreal around its design creativity." With this project, our North American metropolis is given the mission to promote structuring actions (competition, public dialogs, etc.) to imagine and "better make the city of the 21th century with more designers." The *YUL/MTL: Moving Landscapes* project is part of it!

"The urban design ideas competition or the landscape creation workshops do not represent the final objectives of a planning process; their sole purpose is to illustrate the collective visions or aspirations of a territory. However, from these ideas, reflected through images and words, it is possible to identify the essential design guidelines to inform the terms of reference of a project. Therefore, the ideation exercises are support tools to define community issues and plan collaborative territory projects."

Introduction

Over the past few decades, the future of cities and metropolises is a concern that is always brought up in major international debates (ex.: World Urban Forum - WUF[1]), as well as in multiple planning and urban project experiences (ex.: Europan[2]; Greater Paris[3]). This trend that contributes to the city reinvention process is expressed, among other aspects, through the willingness of several cities around the world to link the redeployment of road infrastructures closely with new urban development strategies. In this context, redefining road and highway infrastructures, in an autonomous and closed-in manner, seems to belong more and more to the past. In this century, this concern implies a change of perspective favoring the implementation of resolutely urban projects. Because beyond the infrastructure right-of-way, it is not emptiness, the desert, or no case to answer. Beyond the right-of-way, the city unfolds, which implies that the infrastructure must be permeable to it.

In our contemporaneity, urban agglomerations are more and more being rebuilt on themselves, with special attention paid to the structuring scope of road corridors as a vector of new urban quality; urban mobility becomes one of the major challenges of the sustainable development of cities. Traffic is no longer only taken into account as a simple distinction between the transportation of people and goods, but in a plural manner. Ways of transportation are juxtaposed. Collective transportation (bus, shuttle, tramway) coexists with cars, cyclists, and pedestrians. They deliver various temporal rhythms in an environment where cars no longer only express speed, but in some ways slowness, as demonstrated by the plethora of traffic appeasement measures used primarily in western cities.

---

1. UN-Habitat (United Nations Human Settlement Program)
2. Europan is a competition program in architecture and urban planning, simultaneously organized in several European countries every two years. Intended for practitioners younger than forty, the competitions have strongly contributed to renewing architectural practices in Europe whilst nurturing a research program on cities and habitats. Europan Europe (online) http://www.europan-europe.eu/fr/ (Page visited on May 6 2014)
3. In reference to the international consultation of ten architecture, urban planning, and landscape firms, held to define the metropolitan region of Greater Paris, to identify its development stakes, and to guide its future. Atelier International du Grand Paris (online) http://www.ateliergrandparis.fr/ (Page visited on May 6 2014)

Figure 0.1: Redefining the western entrance to the city of Brest through implementing a tramway. (Photo credit: Yves Poullaouec-Gonidec, 2014)

The Invention of Urban Mobility

Urban mobility is thus evolving, and several European cities (Figure 0.1) have recently been redefining their major roads and entranceways by developing public transit at the expense of private cars. From this perspective, population movements for daily activities associated with work, shopping, and leisure imply considering transportation axes as real public spaces that integrate all adjacent territorial components. Boundaries between the pavements (right-of-ways), public spaces, and habitats are intentionally blurred to create an urban ensemble, an urban coherence that could be qualified as landscape scenography[4].

Thus, moving around a city becomes an experience focused on qualities that are appreciated by a slow and measured discovery of places. In fact, users moving around are placed in such a way as to experience the city from a new angle through which the slow side scrolling offers an easy reading of places. According to this perspective, axial quality and speed are no longer paramount to the moving experience. The

---

4. The question of "landscape scenography" clearly directs to the classic definition of the term "landscapes": either a territory which offers a view to the public and which is subject to interpretation, or the experience of an individual or a community.

Figure 0.2: The half-underground development of the Ronda Litoral, a fast lane located between the Gothic Quarter and the Port of Barcelona (Spain), allows the creation of new public spaces on the surface. (Photo credit: Sylvain Paquette, 2011)

crossed urban contexts thus become the key elements of this experience, showing us that "landscape mobility"[5] is happening in cities of the twenty-first century.

Parallel to the radical urban transformations that change the perspective of a travel experience, the question of accessing highway corridors remains one of the challenges of urban transportation infrastructure. Several projects carried out over the last few years reflect a willingness to rethink these large infrastructures. Again, we are witnessing major transformation and change in perspectives regarding mobility.

The redevelopment of fast lanes that pass through central districts of Barcelona (known as the "Rondas") is one of the most well-known illustrations of this evolution.[6] In addition to winning back adjacent neighborhoods (Figure 0.2), Joan Busquets, Director of the planning department of Barcelona during the 1980s, also considers the road and

---

5. By "landscape mobility", we mean a vehicular movement that offers a landscape perspective, providing a view from staging a traveled distance.
6. Gourdon, J. -L., C. Werquin and A. Demangeon, 2000. *Boulevards, rondas, parkways…des concepts de voies urbaines*. Centre d'études sur les réseaux, les transports, l'urbanisme et les constructions publiques (CERTU), Lyon.

Figure 0.3: The *Buffalo Bayou Promenade* project (Houston, United States) offers a new urban vitality for this local corridor, which is in close interaction with the highway infrastructure. (Photo credit: SWA Group, 2014)

highway network as a real "visual material of the city" because[7] it gets erased in the soil and restores urban viewpoints and in other cases heightens infrastructures that offer fascinating urban skylines.

Elsewhere, emphasis is placed on developing green corridors. The "Green Carpet Plan[8]", put in place in the wake of burying the urban portion of the A2 highway in Maastricht (The Netherlands), aims to provide better integration of road infrastructure in the city. Different planning principles were suggested – for example, the design of a green recreational promenade around the city from north to south, and the implementation of a cyclist or pedestrian lane to open up targeted urban areas or provide access to green areas. In Houston, Texas, redefining the *Buffalo Bayou* local corridor was made possible through the creation of a linear park, which allowed the reuse of land located under the highway infrastructures (Figure 0.3). In doing so, this project enhances the experience for motorists and other commuters entering the city by giving a new transportation capacity to a local corridor.[9]

---

7. Ibid., p. 85.
8. A2 Maastrict (online) http://www.a2maastricht.nl/nl/dp/english.aspx (Page visited on May 6 2014) and West 8, Groene Loper - A2 Maastricht (online) http://www.west8.nl/projects/groene_loper_a2_maastricht/ (Page visited on May 6 2014)
9. Hung, Y.-Y. and SWA Group, 2011. *Landscape infrastructure. Case studies by SWA*, Birkhäuser, Basel.

Figure 0.4: Aerial view of the Yan'an highway (Shanghai, China) and the development project of the Yan'an Zhong Lu Park taken by the WAA inc. firm. (Photo credit: Vincent Asselin)

Through their desire to consolidate a network of green areas through highways, examples of projects similarly carried out in Shanghai (Figure 0.4) magnify the aerial infrastructure with night lights and develop a major public park at its base.

In Melbourne, Australia, along the highway axis that connects the international airport and downtown, a different strategy was adopted. A more formal, artistic approach allowed marking the entrance to the city while emphasizing abstraction and excess in terms of scale. Different installations were inserted, including one with thirty-nine red sticks measuring 100 feet in height placed along the airport expressway.[10] This example illustrates the willingness to use (or even disguise) highway right-of-ways to give meaning to one's journey. Note, however, that this management strategy does not necessarily challenge the territory. The meaning of the intervention is primarily linked to the idea of creating a collection of items along the highway, like a gallery of artistic curiosities with no territorial affiliation.

In contrast, the new project of French landscape architect Bernard Lassus, for the construction of a portion of the E6 highway in Sweden

---

10. Beck, H. and J. Cooper, 2000. *Denton, Corker, Marshall: rule playing and the ratbag element.* Birkhäuser, Basel.

(Figure 0.5), provides another perspective regarding the highway system. It tells the story of a journey through a country – a story that dates back to the Stone Age.[11] Here, it is about recognizing the historic (and prehistoric) depth of a route and evoking "the link between today's travelers and our ancestors"[12], achieved by placing sculptures along highways between Vellinge and the port of Trelleborg in southern Sweden. Lassus advocates the idea that the highway can become a landmark as a technical object. To support his claim, he identifies three themes (reindeer horns, prehistoric arrows, and chariots and trucks wheels) that evoke the depth of the territory and its appropriations.

> "The automobile movement allows a kinetic approach in going forward and backward in time through the three symbols. Each indicates the period referred to at this point of the journey"[13]

Through this landscape evocation, Lassus invites the highway traveler to interpret the visited places. The highway is not a trivial experience. It is a place where a narrative story lives for the duration of a trip. In doing so, the approach is resolutely based on landscape through the territorially evocative experience [14].

Closer to home, in the city of Quebec, an initiative of the ministère des Transports du Québec and the Commission de la Capitale nationale du Québec is committed to redefining one of the major roads that provides access to the city center, the Champlain Promenade, so that it becomes an iconic entrance to this tourist area. At the heart of this strategy is the protection and enhancement of the spectacular views of coastal landscapes presented by this corridor (Figure 0.6). These interventions were made following extensive discussions initiated in the 1990s, which helped to characterize all road access routes to the capital in terms of the quality of generated landscape experiences[15]. The interest of this project lies in the very name of the road. The promenade is the basis of the approach, undertaken to achieve a wandering road that highlights the banks of the Saint Lawrence River and, more broadly, the upper town's low terrace. Here, coastal landmarks unite

---

11. Based on archaeological studies, the designer evokes the idea of a path once taken by reindeer and mammoths who rode northward.
12. Niordson, H. and T. Erlandson in Lassus, B., 2014, *Évocations de la longue histoire de l' autoroute Européenne E6*.
13. Lassus, B., 2014, *Évocations de la longue histoire de l' autoroute Européenne E6*.
14. Note that for over twenty years, Lassus has developed a deep thinking on highway landscaping in Europe through layout planning. He contributed greatly to make the transportation equipment be a real landscape project that makes the territory look like a constantly moving street lamp, but that also provides information on identity, quality and attractions of the visited places. See Lassus, B., 1998. *The Landscape Approach*. University of Pennsylvania Press, Philadelphia.
15. For reference see St-Denis, B., C. Marcoux, and M.-C. Paradis, 2003 as well as Jacobs, P. et coll., 1998.

with recreational facilities (parks, flowerbeds, bike lanes, trails, etc.) that reaffirm the place's identity through their expression. As such, the road-redefining project brings about and expresses the idea of landscape.

This quick overview of road and highway corridor requalification projects demonstrates the importance that we give today, internationally, to the quality of road equipment and the new roles and uses we give to them, but more broadly, to the quality of crossed landscapes, especially along city entrance routes. This change primarily reflects a desire to bring the experience to the heart of a commuter's journey, as well as to associate an infrastructure with its territory.

Such a foundation obviously involves reconsidering the way projects are implemented, including the planning process. In this regard, the United States' Context Sensitive Solutions (CSS)[16] set a benchmark. These solutions refer to the consultation and decision-making process regarding transportation, aiming to take into account the surrounding context (e.g., environmental, aesthetic, historical or social) of all aspects of planning and all stages of project development.

The CSS approach aims to develop transportation infrastructures based on collaboration, interdisciplinarity, and the participation of civil society. Among the key principles of this approach are the desires to understand the contexts of implementation ahead of the project, to establish a project vision shared by various stakeholders while maintaining continuous communication with them, and to make creative and flexible solutions possible for transportation that simultaneously preserve and enhance community environments and the characteristics of the natural environment[17].

Finally, design solutions based on context add value in terms of the quality of living environments, insofar as they seek to meet the needs of all users of the transportation corridors, not only users who travel on them, but also users of the surrounding areas.

If these issues are generating new urban planning strategies, as well as instigating innovative projects in some cities, these questions are part of a longer tradition that takes into consideration landscape dimensions in the road transportation field.

---

16. United States Department of Transportation, Federal Highway Administration, 2007. What is CSS? [Online] www.fhwa.dot.gov/context/ (accessed 29-04-2014); Michaelson, J., G, Toth, and R. Espiau, 2008. *Great Corridors, Great Communities: The Quiet Revolution in Transportation Planning*. Project for Public Spaces, New York.
17. Ibid

Figure 0.5: Evocations of the long history of the European E6 highway (in-between the cities of Trelleborg and Vellinge, Sweden), Bernard Lassus, design and landscape architect. (Photo credit: Bernard Lassus, 2013)

Figure 0.6: Redefining project of the Champlain Promenade at the entrance of the city of Quebec (Canada). (Design and implementation: The Daoust Lestage inc. consortium | Williams Asselin Ackaoui | urban planning option. Photo credit: Marc Cramer, 2008)

As places of landscape experiences, the road and highway become linear panoramic viewpoints that allow us to assess territories' meaning.

The premises of a North American preoccupation

The concept of a "parkway" or "scenic route", implemented at the end of the nineteenth century in the United States, represents an important initial moment of reflection involving road and landscape. At the time, these routes already testified for a desire to enhance landscape features that the infrastructure went through (e.g. forest quality, rural setting) and the need to establish high quality civil engineering structures. Although certain road segments drawn from this concept are today designated as national historic sites in the United States, like the *Merritt Parkway*[18] that crosses part of Connecticut, this road infrastructure model is not widespread in the twentieth century.

In 1964, the pioneering publication "The View from the Road" was not foreign to the comeback of this concern. The authors therefore bring up the importance of the visual experience of driving:

> "Since the highway experience is one of continuous forward-directed movement, the approach and attainment of successive goals is an important feature of it. In addition, the driver seeks to gain some sense of the structure of the surrounding environment, outside the road itself."[19]

The book suggests reading the visual expression of landscapes while driving, a concept which was innovative at the time, and still influences several highway landscape visual analysis methods developed by North American government agencies[20]. The concepts of landmarks, visual

---

18. Radde, B., 1993. *The Merritt Parkway*. Yale University Press, New Haven.
19. Appleyard, D., K. Lynch and J. Myer, 1964. *The view from the road*. M.I.T Press, Cambridge, p. 24.
20. Paquette, S., P. Poullaouec-Gonidec and G. Domon, 2009. *Quebec Landscape Management Guide: Reading, Understanding, and Enhancing the Landscape*, Ministère de la Culture, des Communications et de la Condition féminine, Quebec; Poullaouec-Gonidec, P. and S. Paquette, 2011. *Montréal en paysages*, Presses de l'Université de Montréal, Montreal.

**Appreciation of the highway landscape
benefits as much from the perception
of the infrastructure and its right-of-way
as from the views towards
the crossed territories.**

openness, and rhythm, to name only a few, are part of the new vocabulary adopted by these agencies' experts. Highways do assist with the landscape experience of the territories they cross, but the book sheds light on the difficult territorial articulation of highway infrastructure on the outskirts of a city:

> "Even if the general images of the city and the highway have been clarified and their interrelation established, there still remains the difficult task of linking the road to its immediate environs. This is most crucial where the driver is about to make the transition to the local landscape of streets and buildings. The highway and the city street are two separate words, mysteriously connected, and coming off the ramp of a modern highway is usually a moment of severe disorientation."[21]

In doing so, it set the foundation for a reflection on the urban landscape enhancement conditions of city entrance roads. Facing design logics of road equipment that primarily rely on infrastructure capacity, traffic flow, and road safety, some recognize that the user experience can not be limited to one mobility goal, that is to say, a simple trip to reach a destination.

> As argued by Bishop: "To take a drive today between the countryside and the city, between the airport and the downtown, or between one suburb and the next is an invitation for an assault on our aesthetic sensibilities and our mental health."[22]

---

21. Appleyard, D., Lynch, K. and Myer, J., op. cit.
22. Bishop, K. R., 1989. *Designing urban corridors*, Planning Advisory Service report no. 418, American Planning Association, Chicago, 1.

The obsolescence and disuse of numerous highway infrastructures initiate an open and creative reflection to reinvent landscapes.

Thus, it becomes essential, he says, to rethink the design principles of urban corridors that act as entrances to cities, a concept he associated with the identity of these corridors. Although this concept of city entrance was not experienced in the United States to the extent it was over the last few years in Europe, reflections coming out of the 1960s contributed to recognizing road and highway infrastructures as the foundation to the discovery of landscapes and their enhancement. By extension, roads and highways will become the new linear viewpoints of our modern world.

On the European side, this concern seems to be primarily associated with the rapid, uncontrolled urbanization of metropolitan areas. This would contribute to the deterioration of peripheral areas, as well as to the heterogeneity and discontinuous nature of the urban context. Local identities would gradually disappear, giving way to certain disorder. Road logic creates low-quality city entrances and highway accesses or urban waste, unrelated to the surrounding neighborhoods. This road logic must be accompanied by a logical development. In this context, city entrance requalification tools were developed, in France for example, based on this urban planning approach[23]. While the North American approach initially considers the user of these city entrances primarily as a "contemplative" consumer of the road decor, the most recent perspectives aim to take into account waterfront views that occupy adjacent urban spaces.

Within debates over the inner workings of the "mechanical city", concerned with the proper functioning of infrastructures on the technical and operational levels, "livable city" issues are increasingly

---

23. Gariepy, M., P. Lewis., N. Valois and L. Desjardins, 2006. *Le cadrage paysager des entrées routières de Montréal*. Ministère des Transports du Québec.

present. Thus, the development logics of these crossed areas, which are also living areas for local residents, invite reflection on all involved stakeholders, such as local authority representatives, development professionals, managers of road infrastructure projects, and citizen representatives[24].

Although a majority of road infrastructure projects still adhere to a sectorial approach, a substantive change occurs in instances where urban planning issues remain relatively marginal compared to concerns of transportation. Lessons to be learned from the above-mentioned examples illustrate the magnitude of the challenges to overcome to fill the qualitative gap between mere technical infrastructure project and a genuine planning and landscape enhancement process. If these challenges are still dreams to be realized today, they were the impetus for the Montreal initiative that is the subject of this book.

YUL/MTL: Moving Landscapes, **from infrastructure to territory**

Within the context of the above-mentioned concerns, the *YUL/MTL: Moving Landscapes* project was used to develop a planning vision of the Autoroute 20 gateway corridor between Montreal-Trudeau Airport and downtown Montreal. This project is based on two main foundations. The first comes from defining highways and roads as unique places of experiences and landscape observation, as well as certain types of territorial promotion. The second considers the nature of the city entrance corridor from Autoroute 20, which passes through an area largely composed of brownfield and shows many signs of Montreal's industrial history. Landscape issues may arise in terms of conservation and enhancement, but also in terms of development.

Beyond the international, and thus emblematic, status of the seventeen-kilometer road corridor used by many tourists who visit the Montreal area, the interest of the targeted segment has its origins in the degraded conditions of its infrastructure. Ordinary, ugly, anonymous, gray, and unstructured are the most-heard adjectives used to characterize the area. However, Montreal is internationally renowned for its dynamism, openness, creativity, and quality of life.

While the City of Montreal is recognized as a UNESCO City of Design[25] for involving more architects, landscape architects, and urban designers to shape the future of the city, the seventeen-kilometer corridor will be the focus, over the next 20 years, of major public and private investments regarding transportation infrastructures and urban projects, and it is even more urgent to think openly, and creatively about

---

24. Ibid.
25. City of Montreal. Montreal UNESCO City of Design (Online) http://mtlunescodesign.com/ (page visited on May 6, 2014)

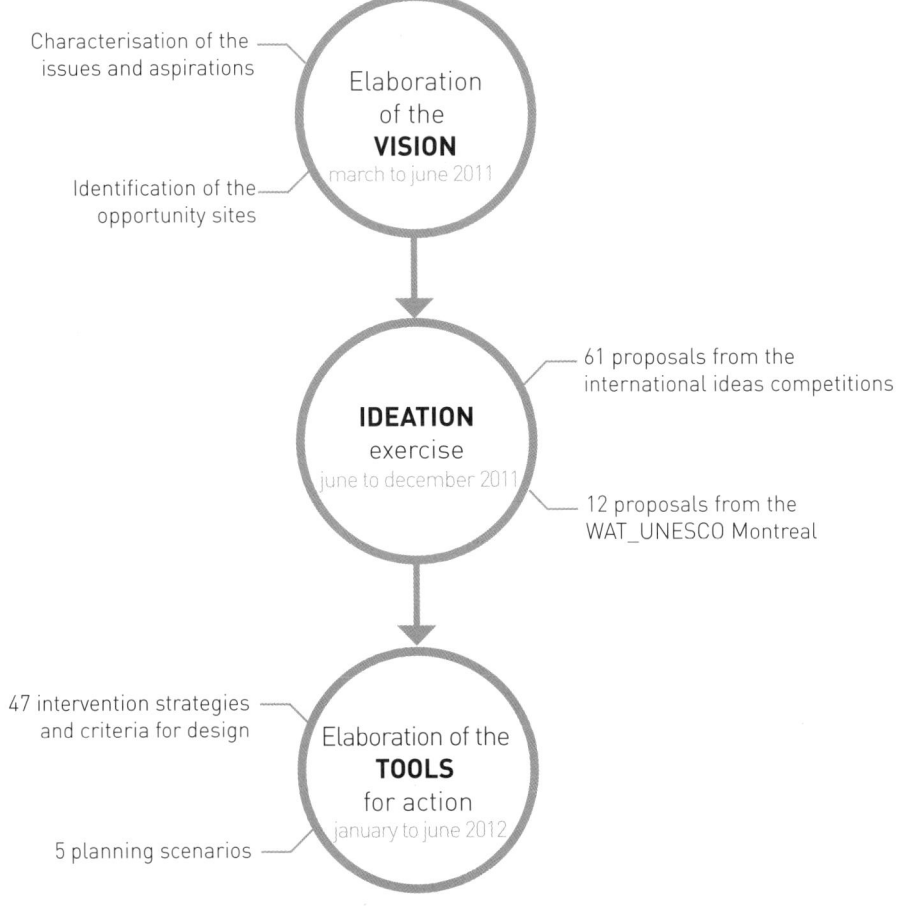

Figure 0.7: Collaborative approach to planning – Autoroute 20 gateway corridor

important development visions to coherently reinvent this entrance area and give it a new urban vitality.

Conducted by the Chair in Landscape and Environmental Design at the University of Montreal (CPEUM) in collaboration with the ministère des Transports du Québec (MTQ), as well as public and private stakeholders gathered around the work table for the Autoroute 20 gateway corridor between the airport and downtown Montreal[26], this initiative takes advantage of a window of opportunity as an emerging, collaborative planning program of primary importance for the future development of the city. It is important to remember that this initiative was also part of the logical continuation of a research program initiated in the early 2000s by CPEUM, also in partnership with the MTQ, on the landscape composition of Montreal's city entrances.[27]

Beyond this first territorial characterization for Montreal, the conceptual foundation of the *YUL/MTL* project is part of a landscape movement, one that understands the redefining project of roads as an opportunity to create landscapes, which directly contribute to the enhancement and development of crossed and lived-in urban areas. It rests upon both recent research on Montreal's urban space[28] and many years of reflection on the issue of landscape, road, and highway infrastructures. Through this positioning, the project is deployed through action-research involving *de facto* researchers and territorial stakeholders.

The planning process of the project is innovative in at least three aspects, which correspond to the main steps involved (Figure 0.7):

> First, it brings together at a work table representatives of all agencies that have a role in planning and development of this area, producing a **coherent and consistent vision** of the development of transportation infrastructures and adjacent areas.
>
> Then, it uses **ideation processes**, such as the international ideas competition and urban design and landscape workshop, to generate proposals that illustrate this vision, to visualize constraints and potential of the territory, and to allow discussion between local stakeholders.

---

26. At the start of the project, the work table was composed of the following private and public stakeholders: Aéroports de Montréal; Agence métropolitaine de transport; Canadian National; Canadian Pacific; Cité de Dorval; Communauté métropolitaine de Montréal; Conférence régionale des élus de Montréal; Parks Canada; Ministère des Transports du Québec; Ministère des Affaires municipales, des Régions et de l'Occupation du territoire; City of Montreal; Côte-des-Neiges – Notre-Dame-de-Grâce Borough; Lachine Borough; Le Sud-Ouest Borough; Ville-Marie Borough; City of Westmount; City of Montreal West.
27. Gariépy et al. op. cit.
28. Poullaouec-Gonidec, P. and S. Paquette, 2011. *Montréal en paysages*, Presses de l'Université de Montréal.

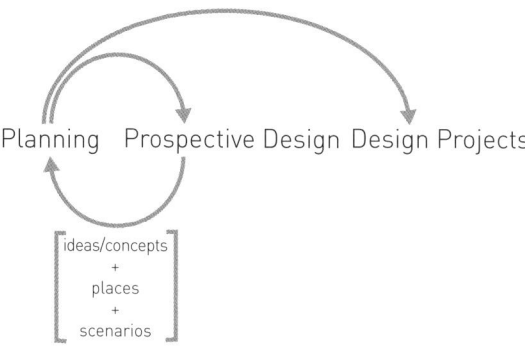

Figure 0.8: Design as a constituent of a planning process (inspired by Vigano, P., 2012).

Finally, it develops **planning tools** in the form of principles and design criteria, as well as planning scenarios that facilitate the appropriation of the process' results and further stimulate exchange among stakeholders.

This approach therefore allows exploring the usefulness of ideation exercises in the planning process (Figure 0.8). In this context, the design is not the final purpose; it is rather used as an essential vision exercise to view the positioning, programming, and implementation of a term of reference. Like the expression used by Paola Vigano, the project becomes a "producer of knowledge"[29]. According to this perspective, it is not the culmination of an approach, but the key to quality processes in urban planning and landscape projects. Ultimately, the success of the collaborative planning process depends on local appropriation of the territorial vision.

Presentation of the book

Several cities around the world today must deal with issues related to the development and redeployment of road transportation infrastructures. The challenge of implementing such projects, while taking advantage of opportunities for conservation, enhancement, and development of landscapes and urban living environments, is considerable. The approach described in this book addresses this challenge. However, approaches that engage a cross-reflection, combining transportation infrastructure, urban landscapes, and

---

29. Vigano, P., 2012. *Les territoires de l'urbanisme, le projet comme producteur de connaissances.* MétisPresses, Geneva.

The ideas competition and the design workshops are key elements of an urban planning quality process.

ideation processes, are rare. Resulting from a single action-research initiative, this publication offers a perspective on emerging issues related to infrastructure landscapes – a perspective that we hope can stimulate the continuing collaborative efforts of coordination and planning undertook in Montreal, and also inspire cities facing similar challenges. As such, this book will benefit both design professionals (architects, landscape architects, urban designers) and urban planning professionals (urban planners, spatial planners, managers), as well as the academic community (professors, researchers, students in the field of urban planning and urban design). It aims to assist the reflection of elected officials and policy makers, as well as all public, semi-public, and private stakeholders of the area, but more broadly, it aims to assist a public interested in the quality of urban living environments, both here and elsewhere.

The sequence of chapters is designed to address the main phases of the process, one after the other. First, a chapter provides an overview of the historical evolution of the entrance corridor of the City of Montreal targeted by the project, while providing cues to understand the socio-political and physical-spatial context at the base of this initiative. The following three chapters present, in order, the singular nature of the process of developing a collective vision, conducted around the work table that brings together the main regional stakeholders, the preferred approaches, and the main results obtained from the ideation processes of the international ideas competition and the urban and landscape design workshop, as well as the resulting planning tools, such as design guidelines and scenarios.

As a conclusion, the book provides some thoughts suggested by the *YUL/MTL: Moving Landscapes* project to face the urgent challenges of redefining highways and roads that are, or will be, facing many urban areas around the world in the upcoming years. It furthermore concludes by stating the relevance of engaging and implementing landscape approaches that are capable of producing meaning to urban areas through the experience of a journey, a meaningful route originating from a place.

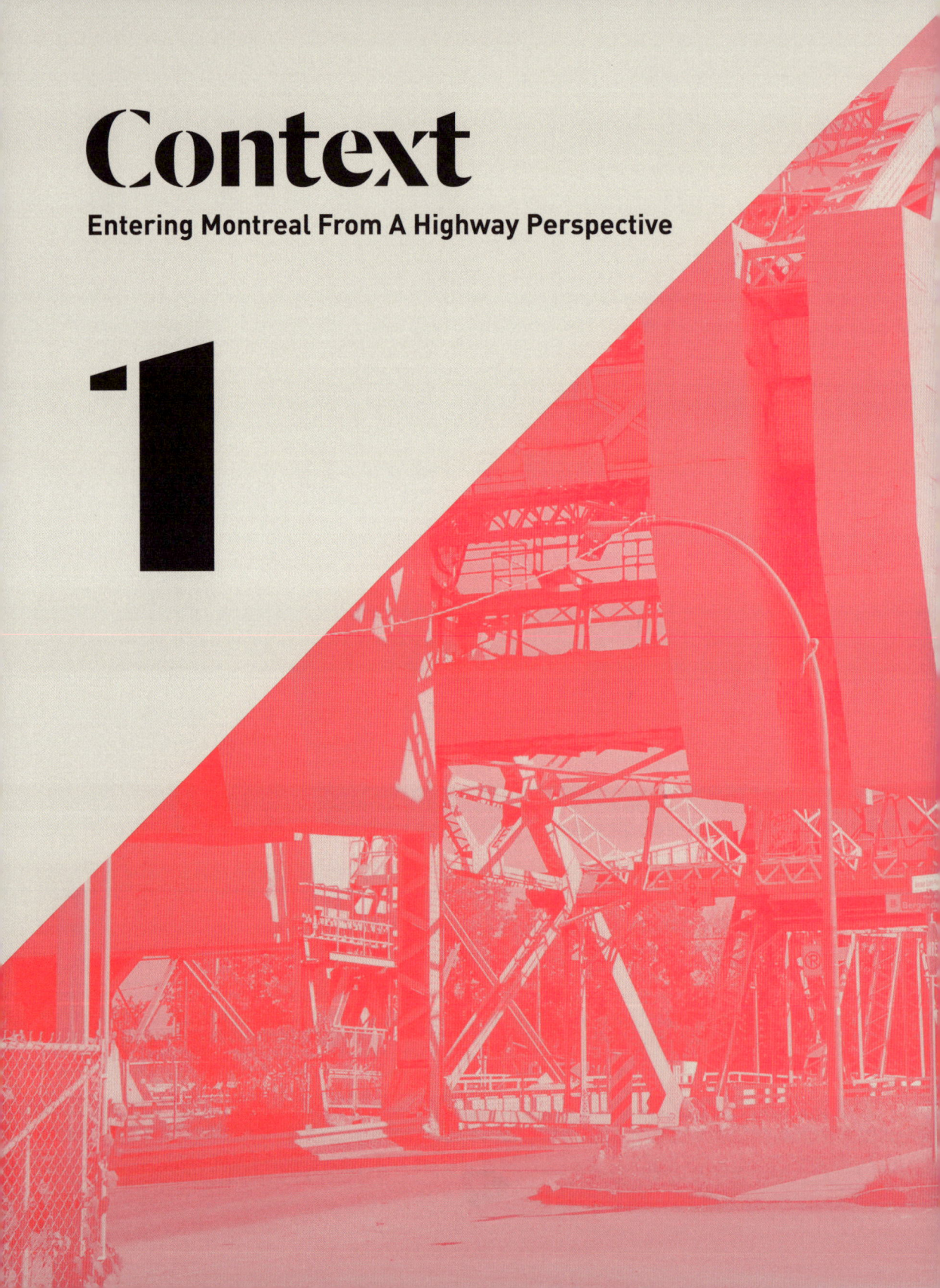

Context

Entering Montreal From A Highway Perspective

1

Far from being only focused on the highway, the notion of a city entrance refers to the whole territory that is in functional or visual interaction with the infrastructure. Similarly, the highway is rarely the only vector of mobility that makes up a city entrance. Composed of highways, roads, railway tracks, or even seaways, a city entrance often constitutes a transportation infrastructure corridor, and reflects a complex process for creating and transforming the territory.

Montreal is an archipelago of the St. Lawrence River. The complexity of the Montreal city entrances consequently roots from a remarkable reality – that of river crossings. The implementation of ferry services, the construction of railway bridges, and then erecting road and highway bridges strongly shaped the organization of the territory around specific development poles and axes. Several railway bridges in the Montreal region have been doubled with road bridges, or have even been transformed into road bridges, generating transportation corridors for local communities and facilitating connections with the rest of the North American continent; the city has long been a focal point in terms of trade.

Montreal is located at the crossroads of three major rivers: St. Lawrence River, that connects the interior of the North American continent to the Atlantic Ocean through the Great Lakes system; Richelieu River, that allows Montreal to reach New York through the Hudson River and Lake Champlain, and the smaller Ottawa River, allowing access to the northern parts of the Quebec and Ontario provinces (Figure 1.1). The location of Montreal at the heart of this maritime system has been, on the strategic side, all the more important because of the presence of the Lachine Rapids, which prevent a continuous navigation toward the western part of the continent. Thus, Montreal was built as a place of transition and exchange, a central commercial intersection for various North American transportation networks.

This chapter aims to depict the city of Montreal's entrances, specifically the Autoroute 20 gateway corridor connecting Montréal-Trudeau Airport (YUL) to downtown Montreal (MTL). It illustrates the complexity of interactions that locally connect the different transportation infrastructures between them and their own territory. The specificity of Montreal's context is put forward in order to appreciate the stakes of the *YUL/MTL: Moving Landscapes* project. The chapter is divided into two main sections. The first traces the history of the deployment

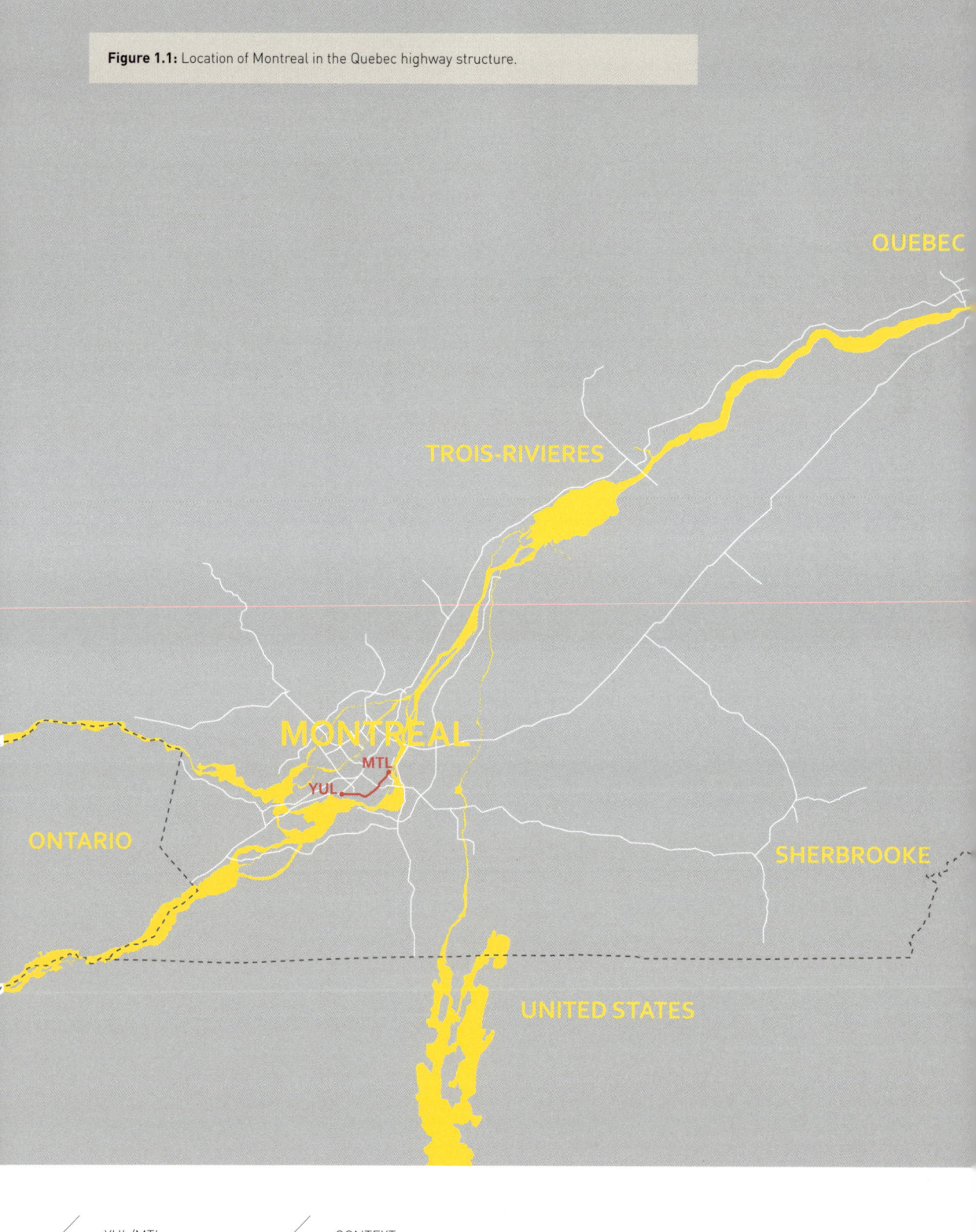
Figure 1.1: Location of Montreal in the Quebec highway structure.

Beyond the purpose of a gate, the city entrance constitutes a corridor that results from a complex infrastructure and urban territories' assembly that unfolds linearly.

of highway infrastructure on the island of Montreal, while stressing the relations between major highways and other transportation infrastructures. The second section locates the various city entrance routes which historically crossed the Autoroute 20 corridor, in order to emphasize the roles played by the succession of different transportation infrastructures in the development, decline, and revitalization of this portion of Montreal's territory.

Deployment of Highway Infrastructures in the Montreal Region

The Quebec highway system was put in place between 1955 and 1975, during a post-war period of strong demographic and economic growth. At the time, Montreal was the metropolis of Canada. It was also during this period that the city hosted the World Exposition of 1967 and the Olympic Games of 1976, two major international events that each generated major infrastructure projects to improve accessibility to the center and its periphery.[1] Subsequently, the development of the highway system is mainly marked by a consolidation of the network. Equipment improvement regarding safety or fluidity also results in targeted operations of road enlargement or interchange reconfiguration.

 The first efforts of highway development on the island of Montreal were primarily intended to facilitate traffic toward the outside of the island. Thus, the first developed highway section, in 1958, is that of Autoroute 15, which goes north toward the Laurentians, a mountainous region renown as a resort destination.[2] The second corridor, established in 1950, is that of Autoroute 40, also called the Metropolitan Highway. It aims to facilitate longitudinal crossing of the island (Figure 1.2). These two corridors do not give direct access to the city center of Montreal. In fact, they are both located on the north side of Mount Royal, while the city center is located on the south side of the mountain.

---

1. Lortie, A. (Dir), 2004. *Les années 60, Montréal voit grand*. Canadian Centre for Architecture: Montreal.
2. For historical markers on road infrastructures development in Quebec, see ministère des Transports du Québec. Historical Capsules (online) https://www.mtq.gouv.qc.ca/portal/page/portal/100ans/capsules_historiques (Page visited on May 5, 2014)

The Autoroute 20 gateway corridor is the only city entrance of Montreal with no river crossing.

Major highway construction sites, started at the dawn of the 1967 World Exposition, aimed at correcting this situation by creating more direct access to the city center, while adding traffic axes toward the main population centers of Quebec and the Northeastern United States. Thus, Autoroute 20 was constructed, which goes east toward the City of Quebec and west toward Toronto, as well as the Autoroute 15, which goes toward New York, and Autoroute 10 to Sherbrooke. Autoroute 40 was also extended toward Ottawa to the west and Trois-Rivières to the East.

The construction of these highways required major civil engineering structures, especially for the crossing of the St. Lawrence River. Thus, the Champlain Bridge was built on the west side of the city center to access Autoroute 10 and 15, and the Louis-Hyppolite-Lafontaine Bridge-Tunnel east of the city center to access Autoroute 20. On the Island of Montreal, the Turcot Interchange is the centerpiece of this highway system. Located at the meeting point of Autoroutes 15 and 20, it stands out due to its height and the viewpoints it offers of the city center.

As a result of these projects, the city center benefited from a single direct highway access point, that of Autoroute 10, which goes along the St. Lawrence River and the 1967 Universal Exposition site by way of the Champlain Bridge. The other planned access to the city center was through Autoroute 20. It was meant to cross Old Montreal along the docks of the Old Port, offering a second longitudinal corridor to cross the island. Requiring the demolition of large portions of historic quarters, and cutting Old Montreal off from its historical link with the port, this controversial project did not start before the Universal Exposition. The extension of Autoroute 20 to the city center was announced at the beginning of the 1970s, but with a revised path, which reduced the required demolitions in Montreal Southwest neighborhoods, and which crossed the city center through an underground tunnel located north of Old Montreal. It thus allowed the preservation of the historical relationship of this heritage area with the St. Lawrence River. However, the open trench dug for the construction of the Ville-Marie Tunnel is still apparent in some portions of the city center, and is gradually being sutured as surface-level urban restructuring projects are completed.

During the 1970s, the highway construction projects were mainly focused on improving access to the metropolitan area. The Autoroute 640 corridor was thus gradually put in place on the North Shore of the Montreal region, and the Autoroute 30 corridor was put in place on the South Shore. These two highways could, in the long-term, have their east and west ends meet in order to create a large peripheral highway. The other two corridors developed in the 1970s are those of Autoroute 13 on the west side of the island and Autoroute 25 on the east side. These two highways allow north-south crossings of the Montreal region. However, the complete crossing of the island of Montreal through Autoroute 25 was only completed in 2000.

With the oil crises of the 1970s, as well as a major economic crisis, the 1980s saw a considerable slowdown in highway construction. Only two projects of highway extension were then completed, Autoroute 25 and 30. In addition, highway projects were then subject to an environmental impact assessment process due to the implementation of more rigorous legislation (Figure 1.3).

At the dawn of the twenty-first century, several highways in the region were more than thirty years old. Like several other major North American cities, Montreal's road infrastructure conditions became critical. Several major bridges and interchanges required total reconstruction. In 2006, the collapse of the Concorde overpass in the northern suburbs of Montreal caused five deaths and triggered a great awareness of the region's infrastructure deterioration. In the years that followed, several reconstruction sites were studied and started to meet the needs of transportation infrastructure maintenance.

City Entrances in the Montreal Region

City entrances toward the center of Montreal that result from this highway system are distinguished, inter alia, by two important elements: the length of the path and direct access to the city center.[3] The city entrances that come from the South Shore must cross the St. Lawrence River. These crossings are made of impressive civil engineering structures, some of them emblematic for the city of

---

3. For a study of complete characterization of city entrances in the Montreal region, see Gariepy, M., et al., op. cit.

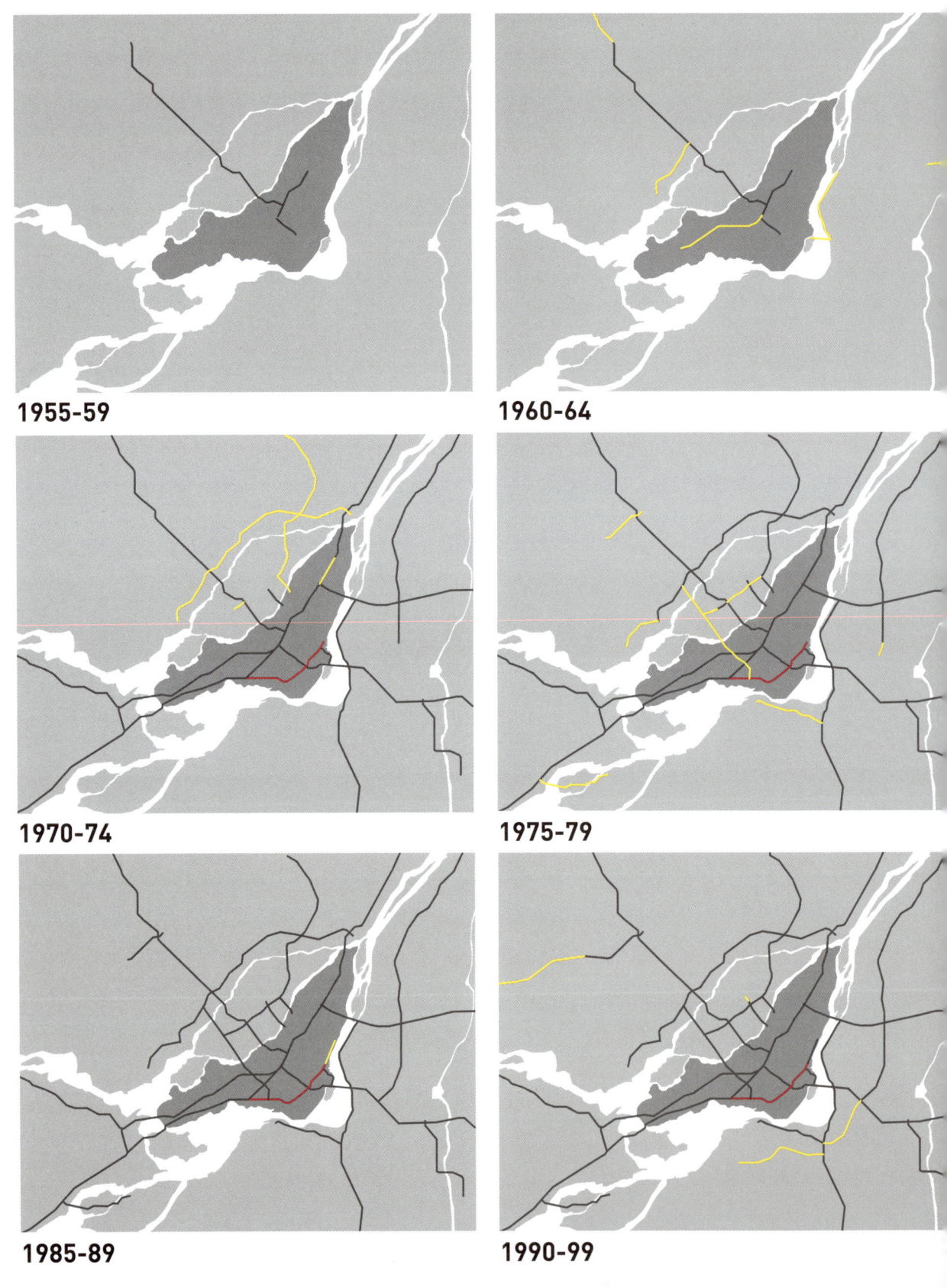

HIGHWAY INFRASTRUCTURE DEPLOYMENT

— YUL-MTL GATEWAY CORRIDOR
— EMERGENCE OF NEW MOTORWAY SECTION
— EXISTING MOTORWAY SECTIONS

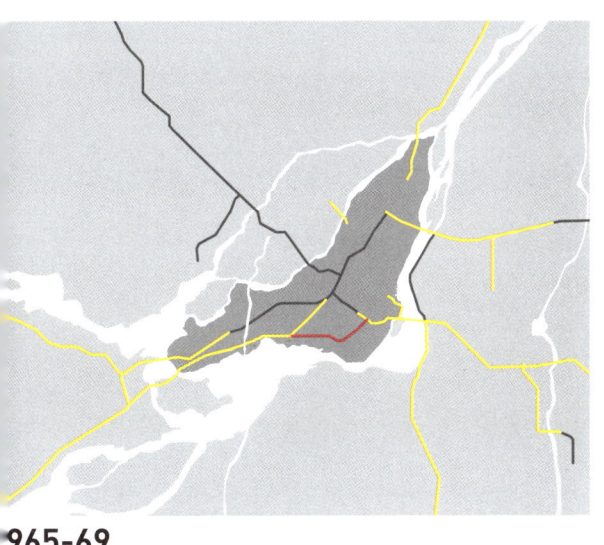

1965-69

Figure 1.2 History of highway infrastructure development in the Montreal region

1980-84

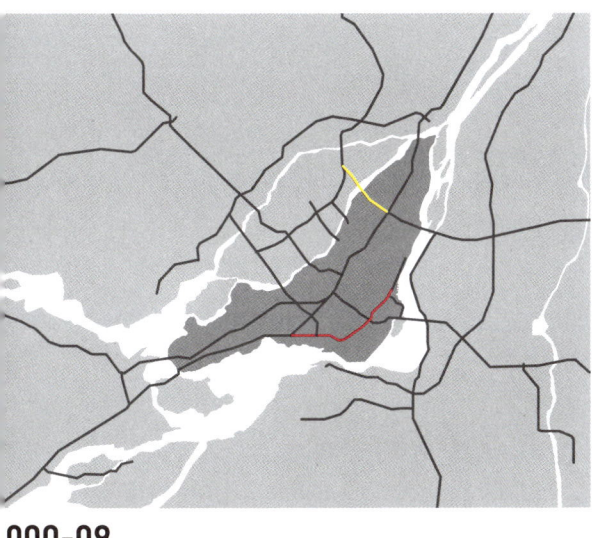

2000-09   2010-13

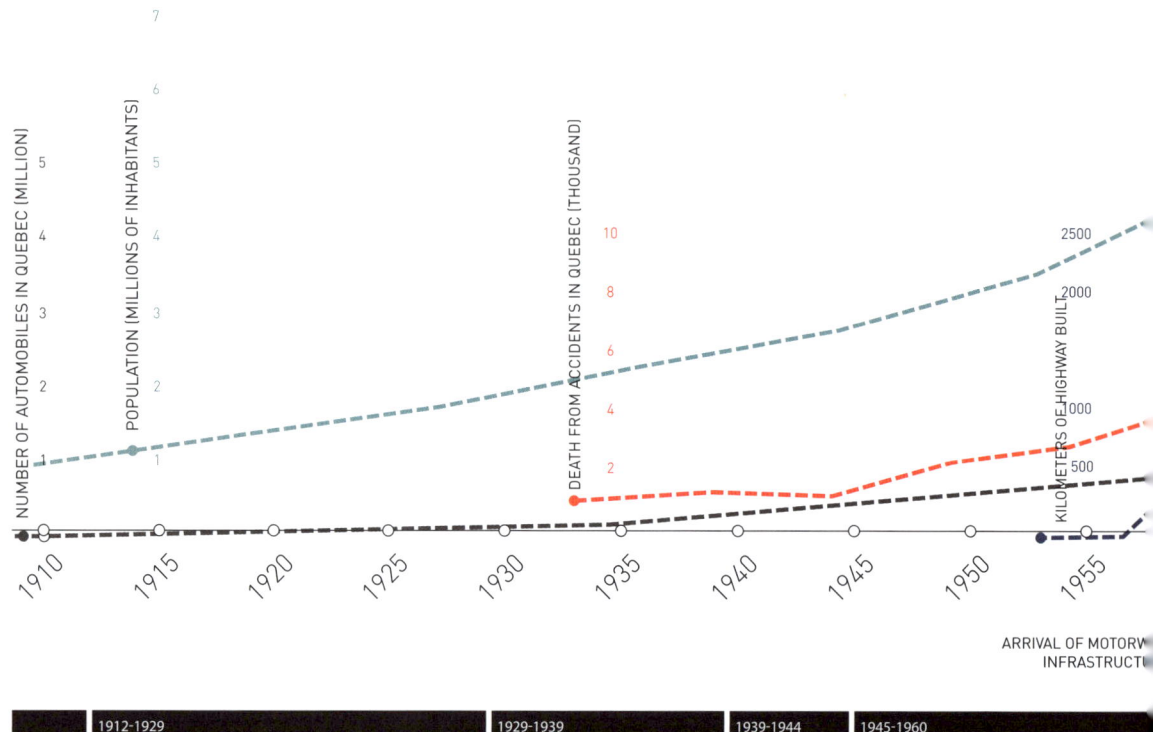

Figure 1.3 Highway infrastructure development context in Quebec[1]

---

1. For statistical and historical data references see Bourbeau, R., 1983, SAAQ, 2012 and Ministère des Transports, Répertoire des autoroutes du Québec, (online) http://www1.mtq.gouv.qc.ca/fr/repertoire_autoroute/autoroute.asp (Page visited on November 6, 2013)

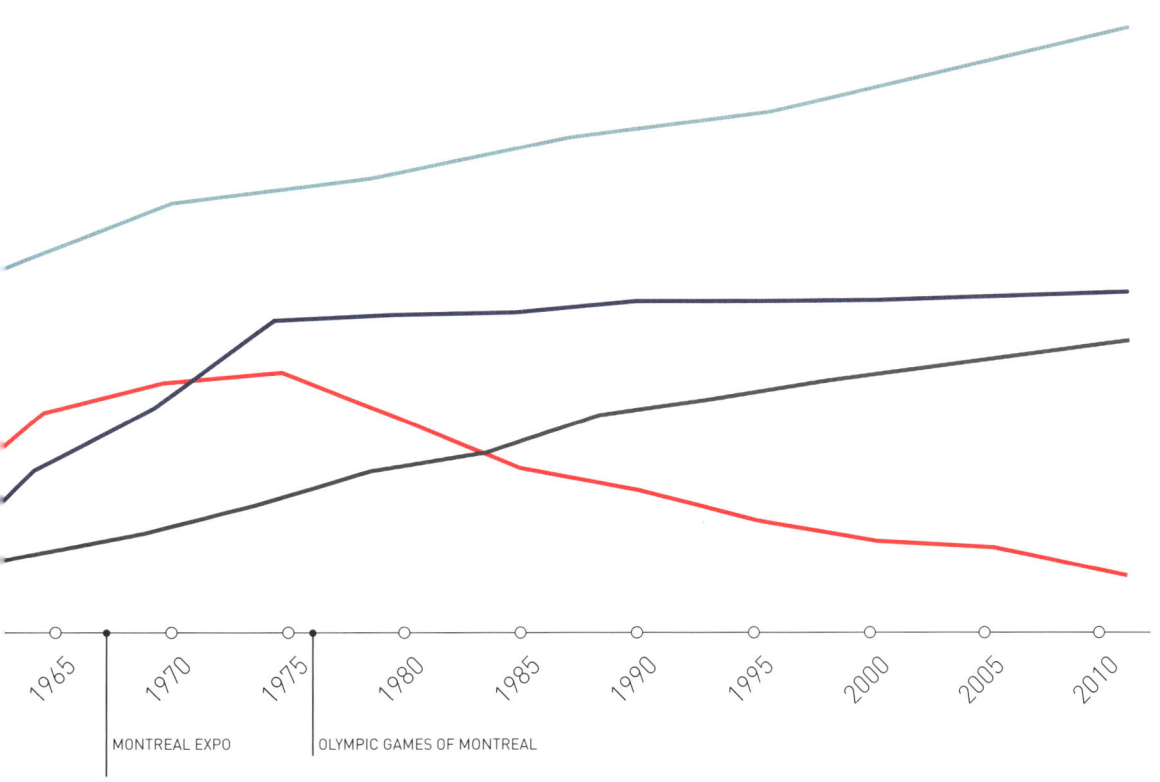

MONTREAL EXPO	OLYMPIC GAMES OF MONTREAL	

1960-1980	1980-1995	1995-2013
DEVELOPMENT OF THE MOTORWAY NETWORK	CONSOLIDATION OF THE INFRASTRUCTURES	UPGRADING OF HIGHWAY INFRASTRUCTURES

1962 Construction begins of Quebec's portion of the Transcanadian Highway and Champlain Bridge
1965 Concorde Brigde
1966 Inauguration of the Montreal Metro
1967 Louis-Hippolyte-La-Fontaine Tunnel
1969 Establishment of the Ministry of Transport
1970 Construction of the Concorde overpass, Laval
1972 Construction begins of Autoroute 20
1972 Construction of Autoroute 720 and Ville-Marie Tunnel
1972 Merger of the Ministry of Transport and Highways + The Environment Quality Act
1973 Government subsidy for public transit
1974 Construction begins of Autoroute 15
1975 Opening of Mirabel airport
1976 Construction begins of Autoroute 40
1978 Construction begins of Autoroute 55
1978 Establishment of the *Bureau d'audiences publiques sur l'environnement* (BAPE)
1979 Land Use Planning and Development Act (LUPDA)

- **1980** Revision of The Environment Quality Act
- **1982** Construction of Viger Tunnel
- **1983** Construction begins of Autoroute 70
- **1986** Systematic visual analysis method, MTQ
- **1991** Construction begins of Autoroute 20 - Bic-Sainte-Luce area
- **1992** Environmental policy of the Ministry of Transport
- **1993** Delegation of responsibility to the municipal sector.

- **1996** Contribution to the Road Network Preservation and Improvement Fund + signage policy
- **2003** Visual monitoring of highway corridor under ecological management (CPEUM)
- **2004** Road and highway landscape analysis method (MEPPRA CPEUM)
- **2006** Transport mode integration support program (PAIM)
- **2006** Sustainable Development Act
- **2007** Action plan of the Ministry of Transport
- **2009** Infrastructure stimulus funds (Federal government)
- **2009** Guide of road projects (MTQ)
- **2000-2013** Aging of structures (viaducts, bridges, interchanges, tunnels)
- **2010** Ground Transportation Networks Fund (FORT)
- **2010** Debate on the future of Champlain Bridge
- **2010-2012** Investment records to improve the quality and safety of infrastructures
- **2012** Northern Plan
- **2012** Metropolitan development Masterplan (PMAD)
- **2012** Opening of the last motorway section of Autoroute 30

Montreal, for the local road network, and for the highway network. Therefore, the highway network articulates the city entrances from the South, mainly around the Champlain Bridge and the Louis-Hippolyte Lafontaine Bridge-Tunnel, the first offering a breathtaking view of the city center (Figure 1.5) and quick access to it. The west, north, and east entrances toward the island of Montreal have shorter river crossings, or are on lower bridges than the south crossings. The result of this is a road experience that is more oriented toward the highway itself, and benefits less from the landscape characteristics viewed from these corridors.

Another element that distinguishes the highway city entrances of Montreal is the more-or-less direct access to the city center. Thus, only city entrances that use the Champlain Bridge (Autoroute 10 and 15), as well as the Autoroute 20 gateway corridor, provide direct access to the center of Montreal. It is noteworthy that crossing the Champlain Bridge means going quickly from the periphery into the city, when several city entrances show a slow progression toward urban centrality.

It is also to be noted that several non-highway bridges allow quick access to the city center, including the Jacques Cartier and Victoria Bridges. Although not considered highways, these two bridges are accessible from the highway system of the South Shore and constitute alternative city entrances to the Champlain Bridge.

The north and east city entrances use the local street network to reach the center, offering closer contact with Montreal daily life. These city entrances go through residential neighborhoods and commercial streets, crossing paths with other road users, particularly pedestrians and cyclists.

Autoroute 20 Gateway Corridor

The Autoroute 20 gateway corridor is different from the others, as it connects to the Montreal-Trudeau International Airport, located on the island of Montreal (Figure 1.4). Therefore, going through the city entrance corridor will not result in crossing the river. In fact, it constitutes a portion of the Autoroute 20 gateway corridor that connects Montreal to Toronto. Nevertheless, the international airport gives a special status to the seventeen kilometers that separates Dorval from the center of Montreal, particularly because of the important traffic and strong visibility to international visitors. With no viewpoint of the river, this corridor allows for an experience of the city's interior.

This axis emphasizes the interdependence of these transportation infrastructures, the presence of one favoring the emergence of another. Thus, far from being only a highway corridor, the Autoroute 20 gateway corridor is also a navigation axis (Lachine Canal) and a railway axis (Canadian National – CN, and Canadian Pacific – CP). Thus, the Autoroute

Figure 1.4: Location of the Autoroute 20 gateway corridor, connecting the Montréal-Trudeau Airport (YUL) to the center of Montreal (MTL).

Figure 1.5: Montreal city center overview from the Champlain Bridge. (Photo credit: Philippe Poullaouec-Gonidec, CPEUM 2014)

20 gateway corridor is a transportation infrastructure corridor of prime importance, given the multiplicity of territorial development opportunities that are made possible, but also a significant constraint, as the network juxtaposition generates the presence of interstitial and residual spaces.

The highway corridor is also at the beginning phase of major changes, both from multiple highway projects undertaken to rehabilitate the infrastructures, and from the great diversity of urban projects that will encourage the development of the territory itself.

The presentation of the historical process of formation and transformation of the Autoroute 20 gateway corridor aims to include the emergence of the highway project in a broader context of territorial development and urbanization. It thus emphasizes that the highway has a double relationship with its own territory: on one hand, the choosing of its path took into consideration the territorial structures already in place at the time of the highway's inception, while on the other hand, the highway itself is a vector for the urbanization and development of the territory.

Then, instead of only focusing on the historic process of implementing the highway between the Montreal-Trudeau International Airport and the city center, the historic portrait seeks to reconstruct the formation of the city entrance corridor since Montreal's founding. Using the city entrance as a device of historical analysis, it is necessary to trace the implementation processes of different modes of transportation that were used to access the city center, identify given points of departure and arrival for each of these modes of transportation, and briefly describe the formal expression of the landscapes.

City Entrance 1: The Shore Path (up to 1847)[4]

At the time of the first human settlements on the Island of Montreal, the Lachine Rapids, located on the St. Lawrence River, obliged any traveler to pause nautical navigation and continue their path on land. A transit corridor was then installed between Old Montreal and Lachine, bypassing the rapids, with the point of return on the waters of the St. Lawrence River.

In addition to the portage along the Saint-Pierre River, the Lachine path, broadly corresponding to the current alignment of Wellington Street and LaSalle Boulevard, seems to be the first access route to the

---

4. Map Source: Bouchette, J., 1831. To his most Excellent Majesty, King William IV. This topographical map of the district of Montreal, Lower Canada (map). 1:175,000, Bibliothèque et Archives nationales du Québec. http://services.banq.qc.ca/sdx/cep/document.xsp?id=0000090116 (page visited on July 10, 2013)

The Autoroute 20 gateway, a profoundly changing territory.

west of the island.[5] As agriculture developed in Montreal, the Upper Lachine Road, built on the heights of the Saint-Jacques Escarpment, offered an alternative to the shore path.

The meeting point of these two paths, located approximately at the mouth of the Lachine Canal, also corresponds to the access point of the St. Lawrence River beyond the rapids. Thus, a crossing system (non-dated) was put in place between the South Shore (Kahnawake) and the Island of Montreal.

Nevertheless, the landscape between these two axes is marked by the predominance of a continuous agricultural environment. Few villages were present between these two points, except perhaps Saint-Henri-des-Tanneries, which formed at the approximate location of the current Turcot Interchange.

City entrance 2: The Lachine Railway - Montreal (1847 - 1883)[6]
In the first half of the nineteenth century, construction of the Lachine Canal significantly transformed the corridor. On the one hand, the canal allows navigation to the doors of Old Montreal. On the other hand, the canal opens a new route that goes along the bottom of the Saint-Jacques Escarpment. This path, also used by Native Americans and explorers who were crossing the island in portage, was not formalized by a road until the nineteenth century, achieved by the creation of

---

5. City of Montreal. Le parcours riverain : Chemin de Lachine et du bord du lac Saint-Louis (online) http://ville.montreal.qc.ca/portal/page?_pageid= 8817,99661605 &_dad=portal&_schema=PORTAL_ (Page visited on May 6, 2014).
6. Map Source: Sitwell, H. S, 1869. Contoured plan of Montreal and its environs, Quebec (map). 1:2 500, Bibliothèques et Archives nationales du Québec. http://services.banq.qc.ca/sdx/cep/document.xsp?id=0000321499 (Page visited on July 10, 2013)

St. Patrick Street on the south side of the canal and the extension of Notre-Dame Street on the north side.

The arrival of the railway also had a significant impact on the city entrance path, in regard to the transportation of people. Although the Lachine Canal offered a direct course to Old Montreal, numerous locks extended the journey time. Thus, the first railway line that was set up on the Island of Montreal was between the Lachine Wharf and the former Bonaventure train station, located near Old Montreal. This railway line was constructed in 1847, and immediately associated with a maritime ferry service that allowed you to reach various places on the St. Louis Lake, such as Kahnawake Wharf, which had a railway line toward the Richelieu River valley, providing the possibility of reaching New York.[7]

Thus, the construction of the Lachine Canal along with the implementation of the railway transformed the way people enter Montreal; the Lachine Wharf became an unavoidable place of intermodal transfer.

At the other end of the line, the creation of the Bonaventure train station testifies to the development of Old Montreal to the north and west. This new station, located outside the city walls, contributed to the establishment of the future business district in the northwest end of Old Montreal, creating a strong polarity and attracting economic activity.

City entrance 3: The suburban trains (1883 - 1925)[8]
Now the property of the Grand Trunk company, the Montreal-Lachine railway was extended in 1883 to reach Dorval, and then, in 1896, the west end of the island of Montreal. In the meantime, the Canadian Pacific Railway (CP) had also established its own railway line, toward the west of the island from Windsor Station. This new line's path went along the top of the Saint-Jacques Escarpment. These two competing lines converged in Lachine to go along the same corridor.

In addition to offering an intercity link between Ottawa and Toronto, the Canadian National Railway (CN) and CP lines also took urbanization to the west of the island of Montreal.[9] Several stations emerged, allowing the development of villages along the shores of the Saint-Louis Lake. It is how Montreal West, Saint-Pierre, LaSalle, Dorval, and other westward cities were born.

Thus, the perceived landscape along the city entrance path changed substantially during this period. In this continuous agricultural

---

7. Hanna, D. B., 1993. *Transport des personnes et développement du territoire de l'agglomération montréalaise : Un essai d'interprétation historique*. Prepared for the Service de la planification du territoire de la Communauté urbaine de Montréal.
8. Map Source: Department of Defense, 1915. Topographic map of Canada, 31-H-05, Lachine (cartographic document) 1:63,360, Bibliothèque et Archives nationales du Québec. http://services.banq.qc.ca/sdx/cep/document.xsp?id=0002684422 (page visited on July 12, 2013)
9. For a historic portrait of the establishment of suburban rail lines in Montreal, see Hanna, D. B., op. cit.

environment, there was a quick formation of a string of villages along the two lines of railway. The city entrance experience was also changed. Instead of taking this path for occasional travel, it became a daily part of the lives of people west of Montreal. It was also the path of the vacationers who populated the shores of Saint-Louis Lake during the summer.

Although points of arrival to the west of the island were increasingly dispersed between multiple stations, the point of arrival to the city center was consolidated. Although the CP added its own station, it was located in the vicinity of the former Bonaventure station, who welcomed the trains of the CN. Thus, the city center entrance occurred in a restricted area, from which travelers and commuters were distributed. Little by little, the area around these two stations became more and more important, as the stations also hosted several other lines of CN and CP; some used the Victoria Bridge, some went to Kahnawake through the Lachine Bridge (a railway bridge juxtaposed to the Honore-Mercier Bridge), and some went toward Saint-Jérôme to the north.

City entrance 4: The National Road 2 (1925 - 1960)[10]
In 1925, the opening of the Galipault Bridge to traffic marked profound changes in the nature of the city entrance path from the west of Montreal, as it reflected the growing importance of the car as a mode of transportation.

The presence of the bridge only provided, at first, vehicular access to the island. The city center access paths remained the same, that is to say, the waterside roadway to Lachine, and either St. Jacques Street (Upper Lachine Road) or LaSalle Boulevard (Lower Lachine Road). Nevertheless, a few adjustments were gradually made to adapt the road conditions to automobile traffic, such as enlarging segments and creating bypass lanes, especially in villages located more to the west of the island, such as Baie-d'Urfe and Beaconsfield[11]. A larger project was being prepared to radically transform the city entrance path: the creation of a large, east-west boulevard across Montreal. Around 1909, various private initiatives inspired by the City Beautiful movement suggested the development of this east-west boulevard, either at the north end of Mount Royal or further south along Ontario Street[12]. In addition to road considerations, these projects were intended to open new territories for real estate speculation.

---

10. Map Source: Department of Mines and Resources, 1940. Topographic map of Canada, 31-H, Montreal (cartographic document). 1:253,440, Bibliothèque et Archives nationales du Québec. http://services.banq.qc.ca/sdx/cep/document.xsp?id=0002670072 (page visited on July 12, 2013)
11. City of Montreal. Le parcours riverain : Chemin de Lachine et du bord du lac Saint-Louis (online) http://ville.montreal.qc.ca/portal/page?_pageid= 8817,99661605 &_dad=portal&_ schema=PORTAL_ (Page visited on May 6, 2014).
12. Noppen, L., 2001. *Du chemin du Roy à la rue Notre-Dame, mémoires et destins d'un axe est-ouest à Montréal*. Ministère des Transports du Québec, Montreal.

To respond to these requests, and ensure public authority regarding the development of Montreal, the provincial government created the Commission of Metropolitan Parks in 1910, "to examine the best means to be taken to create a system of parks, of improved communication channels and model housing for the working classes on the Island of Montreal and on Jesus Island."[13]

Although these ideas on the creation of the east-west boulevard lead, at first, to the creation of the Metropolitan Boulevard on the north side of the island, and therefore setting it outside of the Autoroute 20 gateway corridor, they had a significant impact on highway planning context. They were at the origin of an extramunicipal commission in charge of exploring the metropolitan planning issues. Thus, the provincial government recognized that the planning of the boulevard had to come from consultations among the different municipal and provincial authorities, since the axis crossed several distinct jurisdictions.

The Autoroute 20 gateway corridor was constructed in a progressive way. After the construction of the Galipault Bridge, the first important action was to develop National Road 2, more specifically the Montreal-Toronto Boulevard, the first large boulevard designed for automobiles along the axis of the Autoroute 20 gateway corridor. It connected the west end of the island to the Saint-Jacques Street. The path of this road went along the rail corridor of the CN and CP. This segment of the National Road 2 appears on the topographic maps of the Department of National Defence in 1936.

At the time of its construction, this boulevard still traveled through a primarily agricultural region up to Saint-Pierre, where it crossed urbanized environments. It is important to mention, although it was intended for automobiles, National Road 2 did not have the highway status that Autoroute 20 has today. At the time, the National Road 2 was a boulevard, crossing other streets at grade level. These crossings were rare in rural segments, but regular in urbanized segments. This arrangement allowed a uniform development along the route, not favoring any place for specific access. However, there did not seem to be a concerted effort to form a continuous built front along the boulevard, such as an urban facade.

Despite the linearity of the Montreal-Toronto Boulevard treatment, the development of National Road 2 distinguishes an important point of the course – the intersection with the Côte-de-Liesse Road and Dorval Avenue. From the first moments of the boulevard's development, this intersection was highlighted by the presence of a roundabout, the only construction of this type along the entire route. The roundabout is an important feature of the territory, as it also indicates the entrance to the Dorval Airport, which opened in 1941, just after the boulevard's

---

13. From Lomer Gouin quoted in Noppen, L., op. cit., p. 23.

inauguration. Thus, the Dorval roundabout gradually became a specific point along the Autoroute 20 gateway corridor.

To access the city center of Montreal, National Road 2 went through Saint-Jacques Street up to Westminster Avenue (Montreal West), then diverged toward the north to reach Sherbrooke Street. At this time, Saint-Jacques and Sherbrooke Street were already largely urbanized, and the traveler was plunged into Montreal life. This situation highlights an important difference that occurred in the development of LaSalle Boulevard and Saint-Jacques Street, the first two access paths to Old Montreal. While the presence of the CP line consolidated the role of Saint-Jacques Street in accessing the city center, LaSalle Boulevard lost its importance in the hierarchy of metropolitan lanes. Real estate developments were low in that area until the propagation of automotive transportation in the second half of the twentieth century. Thus, in 1915, the only urbanized sector beyond Verdun was a small area located around LaSalle Station. This small kernel was consolidated by the construction of the first Honoré-Mercier Bridge in 1934.

It is also important to note that, by focusing on Sherbrooke Street, the course of the National Road 2 determined a new entrance point to the city center, more toward the north than it had previously been. This movement of the city center's main entrance also marked the spread of the city center toward the north.

City entrance 5: Autoroute 20 (1960 - 1970)[14]

In 1941, the Planning Department of the City of Montreal was created. Over the 1940s, several studies were produced by this department, to better know the state of buildings and infrastructures, as well as the living conditions of Montrealers. One of the studies that had a considerable impact on the development of the town was a report entitled "An East-West Autostrade"[15], which was published in 1948, proposed to provide the City of Montreal with a highway (autostrade) that would cross the city center from east to west, to relieve the streets from traffic congestion, as well as to better serve the Port of Montreal and the industrial zones on the west side of the island.

This report laid the foundations of what Autoroute 20 was going to be. The suggested course of the autostrade reached the crossroads of Montreal-Toronto Boulevard and Saint-Jacques Street, in Saint-Pierre, to go along the CN railway in Turcot shunting yard, reaching Old Montreal.The report is also interesting for the vision it offers on the form that a highway project should take. It mentions that:

---

14. Map Source: Department of Defense, 1967. Topographic map of Canada, 31-H-05-g, Lachine (cartographic document). 1:25,000, Bibliothèque et Archives nationales du Québec. http://services.banq.qc.ca/sdx/cep/document.xsp?id=0002671938 (Page visited on July 12, 2013)
15. City of Montreal, Planning Department, 1948. Une Autostrade Est-Ouest. Montreal, City of Montreal.

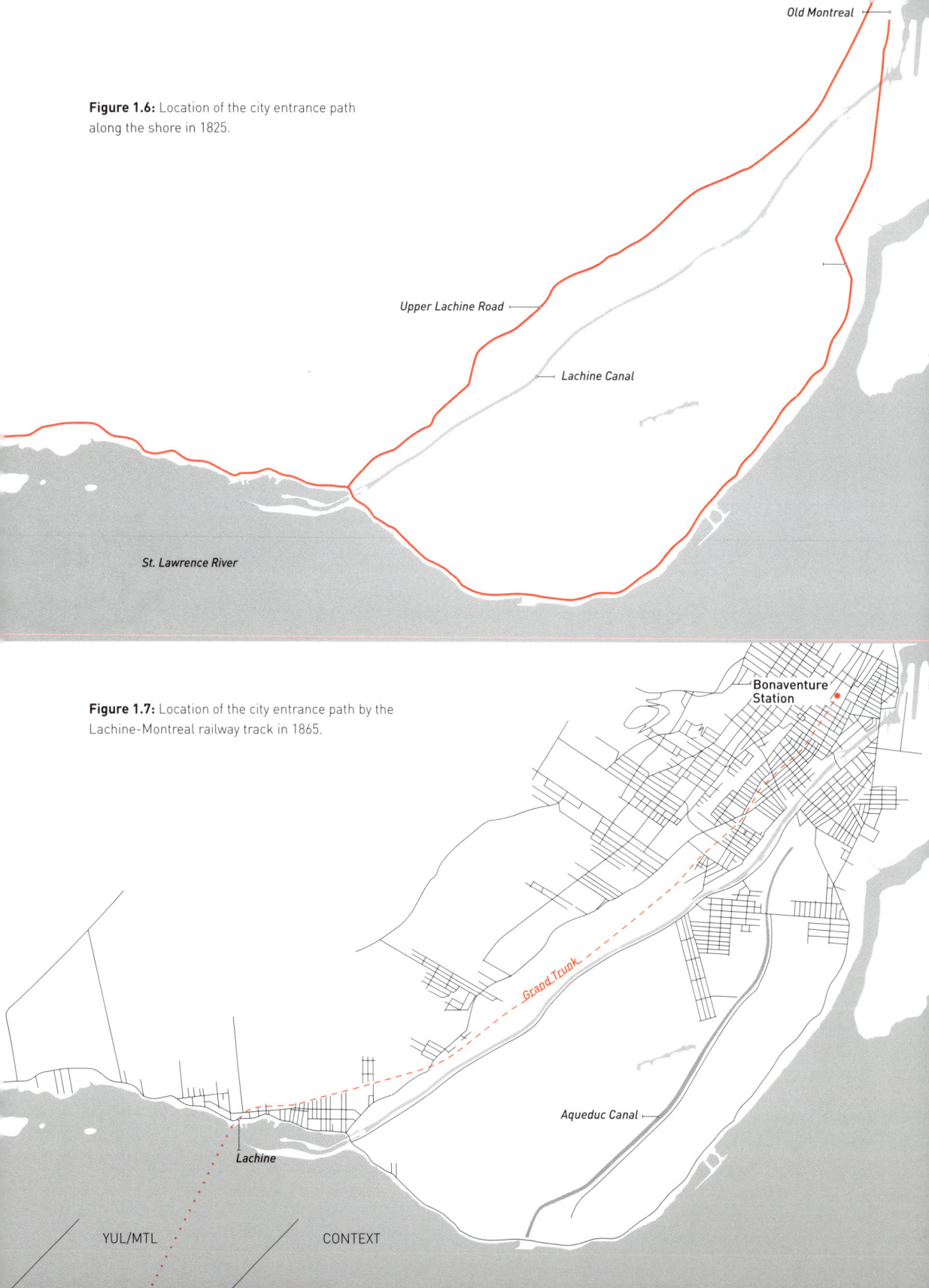

Figure 1.6: Location of the city entrance path along the shore in 1825.

Figure 1.7: Location of the city entrance path by the Lachine-Montreal railway track in 1865.

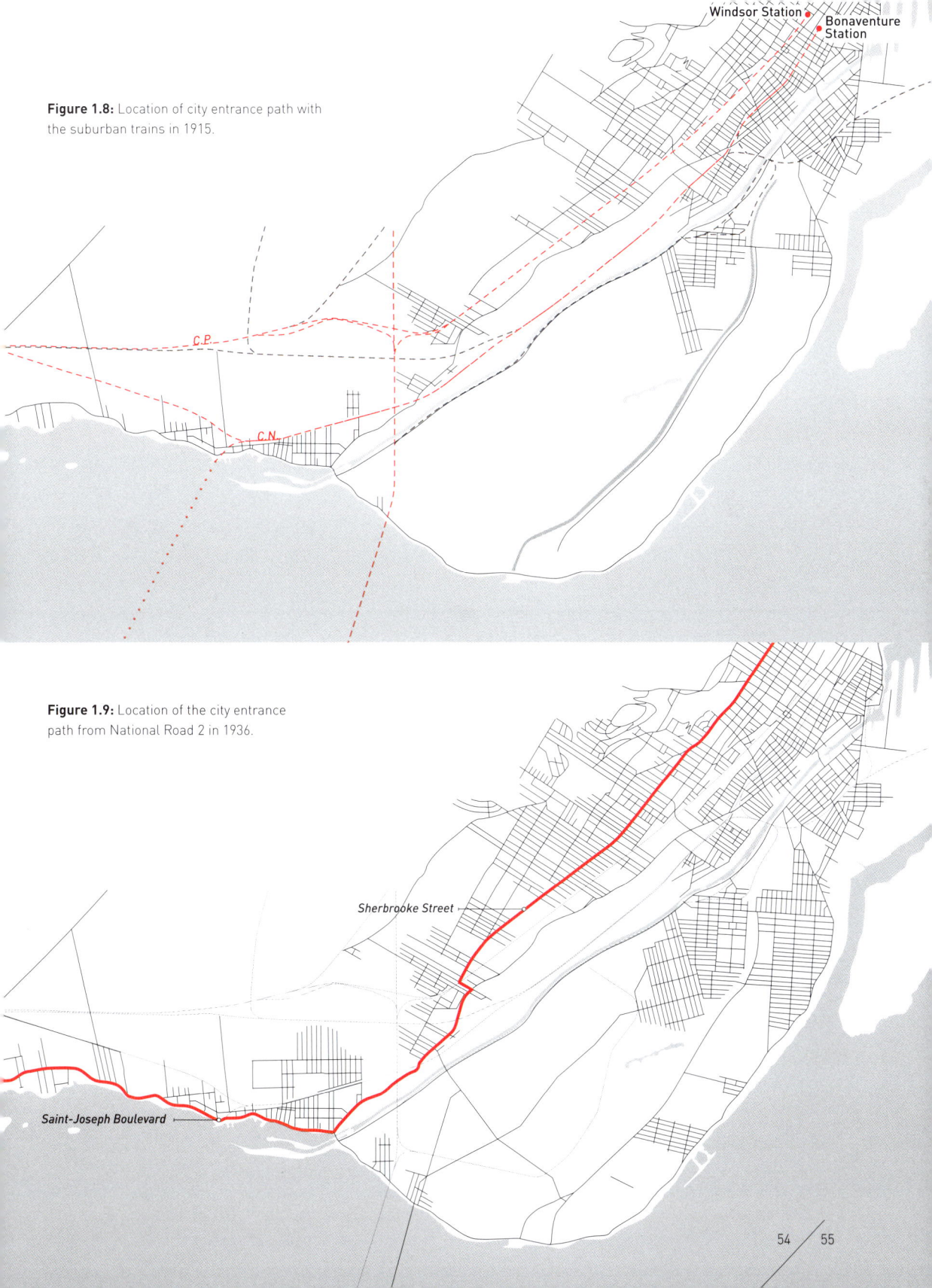

Figure 1.8: Location of city entrance path with the suburban trains in 1915.

Figure 1.9: Location of the city entrance path from National Road 2 in 1936.

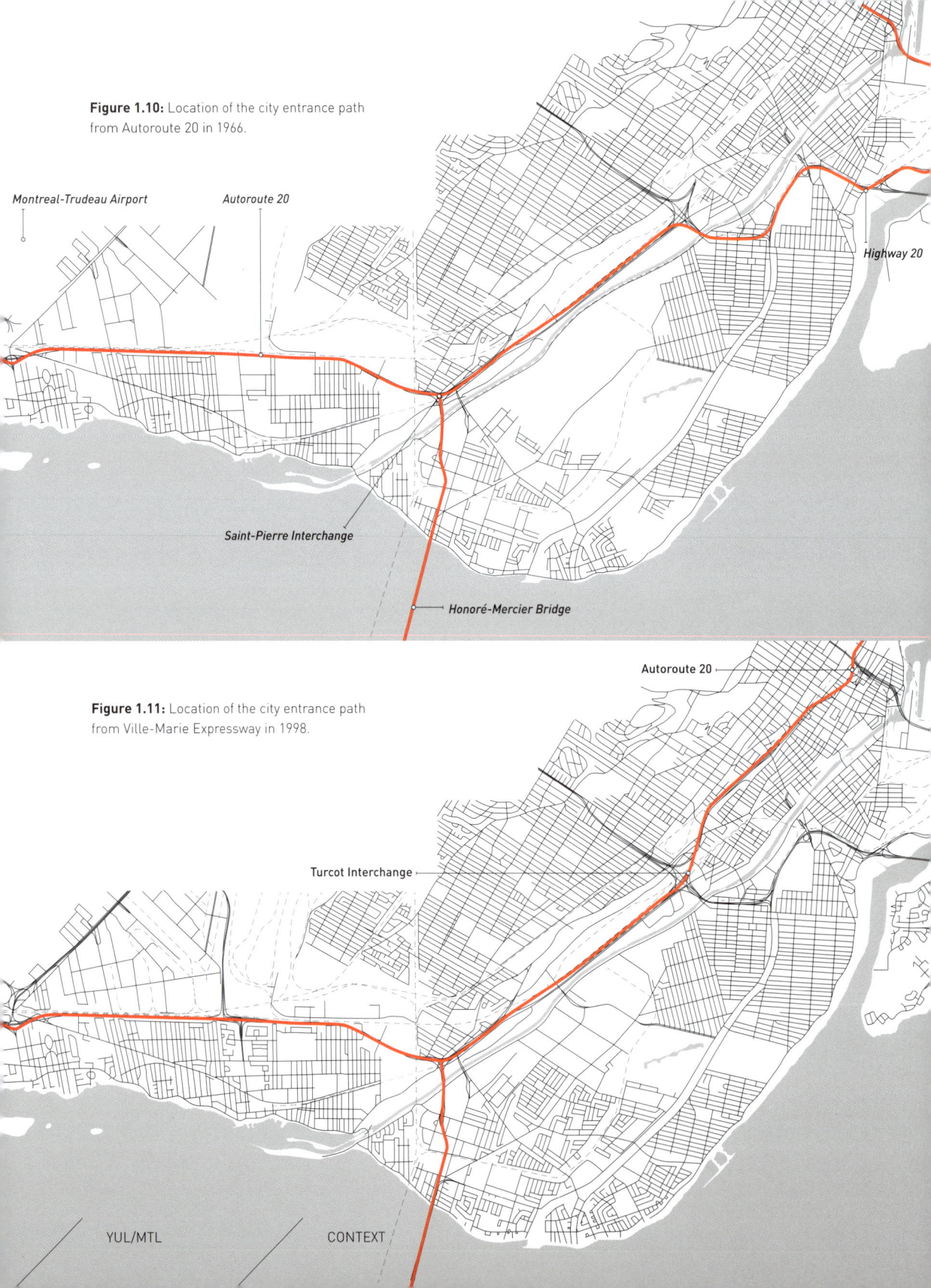

Figure 1.10: Location of the city entrance path from Autoroute 20 in 1966.

Montreal-Trudeau Airport · Autoroute 20 · Highway 20 · Saint-Pierre Interchange · Honoré-Mercier Bridge

Figure 1.11: Location of the city entrance path from Ville-Marie Expressway in 1998.

Autoroute 20 · Turcot Interchange

YUL/MTL   CONTEXT

"In sum, an autostrade is an artery without traffic lights or level crossings in which the tracks in opposite directions are divided, and the entrances and exits, reserved to a few selected locations; cross streets, important intersections, are arranged in overpasses or underpasses. It is a track to be measured, especially at the periphery of the region, a continuous movement of traffic, not hampered by departures, stops, park and pedestrian problems and or the risk of an accident is greatly reduced."

"An appropriate mix of segments that are not bordered by bypass or connecting lanes gives the whole course an aspect of park, and if field residues are present in suitable places, we can advantageously set up parks or playgrounds in neighborhoods."

"Although built to expedite the passage of private and commercial vehicles, the autostrades also constitute, in urban areas, an advantage for public transportation by facilitating the movement of trams and buses in the streets on surface. When an autostrade is parallel to the lines of trams, public transportation or a suburban, their tracks can be arranged on a central terrace. [16]"

This vision of the highway illustrates the desired complementarity between the development of fast traffic lanes and the enhancement of the territory, especially by planning parks or public transit routes.

The decision to organize the Expo 67 gave body to this autostrade project. To welcome visitors, the three government levels quickly measured the work to be accomplished and prepared an ambitious plan of highway construction, which the east-west autostrade was part of. [17] Nevertheless, only the western segment of this highway was completed in time for Expo 67. Thus, the Montreal-Toronto Boulevard between the Galipault Bridge and St. Pierre was profoundly restructured, to be transformed into a highway with limited access. Then, the Saint-Pierre Interchange was built, as well as the segment of Autoroute 20 that goes through the heart of Turcot Yard to the Turcot Interchange.

Thus, during Expo 67, the city center access path allowed motorists to cross a mixed environment that was already beginning to resemble what users can see today, which is a territory heavily segregated between residential parts – located on the south side of Autoroute 20 – and the industrialized parts – located on the north side of the highway, as well as in Turcot Yard. Nonetheless, it is interesting to note that between Dorval and Lachine, due to the presence of the rail infrastructure, Autoroute 20 never truly gave access to the industrial

---

16. Ibid., p. 7-8.
17. Lortie, A., op. cit.

areas that it bordered. The historical development maps identify the Côte-de-Liesse Road, which became a highway in 1966, as a way to access this industrial sector. While other highways in the metropolitan area offer an industrial corporate front, Autoroute 20 faces the backyards of the industrial zone, including wasteland and stacks of containers. Only the short segment, in Lachine, offered an industrial front for the highway landscape.

Regarding city center access, the entrance path then required one to follow Autoroute 15 toward the south, and then the Bonaventure Expressway, both newly created before Expo 67.

City entrance 6: 720 Expressway (1970 - 2014)[18]

As the only large missing segment of this road network, the Ville-Marie Expressway (720) gives direct access to the city center and allows its crossing. The east-west autostrade report of 1948 favored a route through the Saint-Henri and the Little Burgundy neighborhoods, between Notre-Dame Street and the Lachine Canal, to reach Old Montreal at Commune Street. This path, going along the shore, was inspired by other North American highway initiatives of the time. Thus, like New York and Chicago, it was planned to give Montreal a highway on the edge of the river.

Although this path had the advantage of servicing the industrial areas of the southwest and Old Montreal, it was challenged. Quickly, objections were raised[19] to avoid the construction of an elevated highway on the edge of the river, particularly that of influential French planner Jacques Greber, whose opinion was requested. Greber suggested transforming the autostrade into an urban boulevard along the Old Montreal, in order to clear the view and to maintain the city-port-river relationship until needs became more important. Finally, it was the proposal of H. P. Daniel van Ginkel and his wife Blanche Lemco-van Ginkel, both pioneers of architectural modernism in Canada, to accommodate the highway next to the rail network of the CP, along the cliff separating Westmount and Saint-Henri, then entrenched in the hollow formed by the St. Martin River to cross the city center, which became the consensus. It was in 1970 that the Robert Bourassa government launched a program of urban demolition and the construction of the Ville-Marie Expressway, which follows this path.

The choice to build the Ville-Marie Expressway on the hillside, on the edge of the railway tracks, offers unique views of the city

---

18. Map Source: Ministère des Ressources naturelles, 1998. Carte topographique du Québec, 31-H-05-200-0202 (cartographic document). 1:20,000, Bibliothèque et Archives nationales Québec. http://services.banq.qc.ca/sdx/cep/document.xsp?id=0002683860 (Page visited on July 12, 2013)

19. For more details on this Montreal urbanisation period, see Legault, G. R., 2002. *La ville qu'on a bâtie, trente ans au service de l'urbanisme et de l'habitation à Montréal*, 1956-1986. Liber, Montreal.

center skyline, as well as of the roofs of the Saint-Henri and Little Burgundy neighborhoods.

This new highway segment proposed two separate entrances to Montreal's city center. The first is located a little west of Guy Street, around the Canadian Centre for Architecture (CCA). Built around the Shaughnessy House which was saved from the demolition program for the highway construction, the Canadian Centre for Architecture also includes the CCA Garden designed by Melvin Charney in the interstitial space between the highway access ramps. This intervention at the edge of the highway led to the creation of an important city center entrance points.

The second city center entrance point is located on de la Montagne Street. This second exit leads to an entrance point of the city center located in the vicinity of the Bonaventure and Windsor railway stations, reminding one of the important role of this Montreal area, as well as the image of the city center. This sector of the city center nevertheless remained largely destructured after highway construction and the relocation of Bonaventure Station.

Autoroute 20 gateway corridor restructuring stakes

Since the Ville-Marie Expressway construction in 1970, the path of the Autoroute 20 gateway corridor hasn't change much. Nevertheless, the presence of the highway generated major restructuring of the territory, particularly because of the gradual disaffection of the railway as a method of transportation for people and goods.

The decrease in railway activities and industrial decommissioning

The loss of vitality of the industrial area (Figure 1.12), which was connected to the CN track in the southwest, was also due to the closing of the Lachine Canal in 1970. The closing occurred due to the opening of the St. Lawrence Seaway, which made using the canal to cross the Lachine Rapids obsolete. Thus, these two factors led to the gradual closure of several industries located along the canal.

Railway tracks that were devoted to the transportation of people were also gradually abandoned[20]. In 1958, the Central New York line abandoned its suburban train service, which used the railway bridge next to the Honoré-Mercier Bridge to reach Châteauguay. CN also gradually abandoned its suburban train service toward the west of Montreal. In 1955, the portion between Dorval and Dorion was interrupted. Then, in 1961, the rest of the line was abandoned. Thus, only CP kept its suburban rail service toward the west of the island,

---

20. Hanna, D. B., op. cit.

Figure 1.12: Industrial wasteland of the city entrance corridor from Highway 20. (Photo Credit: Philippe Poullaouec-Gonidec, 2011)

which is now integrated into the service of the Agence métropolitaine des transports (AMT).

The restructuring of rail transportation services generated several changes in both the southwest borough and the city center of Montreal. On the one hand, from the 1920s, the project of a new Bonaventure Station on the location of the current Central Station made its way. The new central train station finally opened in 1943, and groups the various services of Grand Trunk that became Canadian National (CN) in the meantime . The old Bonaventure train station was then only reserved for the transportation of goods. A fire resulted in the complete demolition of the old train station, with the dismantling of the railway itself freeing a strip of land for urban development.

It is to be noted that receiving trains and passengers at Windsor Station was also abandoned by CP due to the construction of Bell Centre, a large sports arena located over the old railway tracks immediately to the west of Windsor Station. The new Lucien-L'Allier Station, built inside the Bell Centre building, then welcomed the suburban train passengers.

The revitalization of the Lachine Canal

In 1976, shortly after the Lachine Canal was closed to navigation, a cycling lane was created on its surroundings (Figure 1.13). Thus, the linear infrastructure slowly acquired a renewed role in the urban structure of Montreal, that of an important axis for the cycling community. This project of a linear park was realized within the broader framework of a project revitalizing its surroundings, undertaken by the federal government.[21] It was only in 2002, however, that the canal itself was reopened for recreational boating.

The various actions of revitalization of the canal, its right-of-way, and its surroundings generated a profound transformation of the urban landscape of the southwest borough. The industrial image that once ruled the area was transformed by the cleaning of industrial sites and the construction of housing.

The progression of the bicycle as a method to access the city center was also a major transformation. It was consolidated with the revitalization of the western portion of Old Montreal. Changes in this sector of the city, combined with the transformation of access methods, reinforced the perception of Old Montreal along the gateway corridor.

---

21. Parks Canada, 2004. National Historic Site of Canada of the Lachine Canal Master Plan (online). http://www.pc.gc.ca/fra/lhn-nhs/qc/canallachine/docs/plan1.aspx (Page visited on May 5, 2014)

The Turcot Complex reconstruction

The Turcot Interchange was considered, at the time of its construction, a work of art and remarkable engineering (Figure 1.14). It is an artwork that visitors coming from Ontario, the Eastern United States, and, of course, from the Pierre-Elliott Trudeau International Airport see when arriving in Montreal.

Due to the outdated infrastructure that made up the Turcot Complex, the provincial government announced, in 2007, an unprecedented investment for its reconstruction. The choice to reconstruct was particularly privileged, as it allowed the reconfiguration of traffic lanes to improve fluidity and safety, as well as to decrease the height of the structures to ease their long-term maintenance.

This Turcot Complex reconstruction project created an important opposition movement that brought together many elected representatives of the City of Montreal and community organizations.[22] Thus, projects resulting from those opposing the reconstruction of the Turcot Interchange were born to demonstrate the feasibility of alternative options for the redevelopment of the highway infrastructure. These proposals came as much from the municipal administration[23] as the civil society organizations[24]. The public hearing report[25], held within the framework of the environmental impact assessment study of this project, testifies to these various oppositions, which mainly revolve around the relationship between public transportation and highway infrastructure, as well as the integration of the infrastructure into the urban frame.

A Gateway Corridor in Search For Consistency

This portrait of highway infrastructure development in the Montreal region, particularly in the Autoroute 20 gateway corridor, illustrates the great interdependence between each mode of transportation. The establishment of major transportation infrastructures has successively

---

22. Benessaieh, K. Échangeur Turcot : un milliard gaspillé, selon Vision Montréal, *La Presse* (online). (April 4, 2011) http://www.lapresse.ca/actualites/montreal/201104/04/01-4386340-echangeur-turcot-un-milliard-gaspille-selon-vision-montreal.php (Pages visited on May 5, 2014); Bisson, B., 2011. Le Comité de vigilance Turcot conseille d'élaguer le projet, La Presse (online). (April 16) http://www.lapresse.ca/actualites/montreal/201104/16/01-4390502-le-comite-de-vigilance-turcot-conseille-delaguer-le-projet.php (Pages visited on May 5, 2014)
23. Bisson, B., 2010. La Ville propose un échangeur Turcot plus compact, La Presse (online). (April 22) http://www.lapresse.ca/actualites/montreal/201004/22/01-4273008-la-ville-propose-un-echangeur-turcot-plus-compact.php (Pages visited on May 5, 2014)
24. Guthrie, J., 2010. Pour un projet Turcot réduit et plus humain, *Journal Métro* (online). (March 25) http://journalmetro.com/actualites/montreal/33154/pour-un-projet-turcot-reduit-et-plus-humain/ (Pages visited on May 5, 2014)
25. Bureau d'audiences publiques sur l'environnement, 2009. *Projet de reconstruction du complexe Turcot à Montréal, Montréal-Ouest et Westmount. Rapport d'enquête et d'audience publique.* (online) http://www.bape.gouv.qc.ca/sections/rapports/publications/bape262.pdf (Page visited on May 5, 2014)

Figure 1.13: Linear park along the Lachine Canal. (Photo Credit: Philippe Poullaouec-Gonidec, 2011)

contributed to the formation of multimodal city entrance corridors. These corridors, including that of the Montreal airport gateway, however, were never part of a coherent project of urban development. Rather, each segment was constituted along the flow of that development's opportunity and under local pressure. Regarding the Autoroute 20 corridor, the abandonment of the urban project specifically designed for the "autostrade" that incorporated the development of parks and the implantation of public transportation lanes did not help to integrate the highway infrastructure into the territory.

The Turcot Interchange Complex reconstruction represents an opportunity in this sense, if it falls within a context of broader planning that includes the participation of local stakeholders in territorial planning. Similarly, the emergence of a new regional development plan[26] that focuses on consolidating public transportation corridors and intensifying their adjacent territory reveals to be an important source for renewing development and planning perspectives of this territory. The multiple stations that punctuate the two lines of suburban

---

26. Communauté métropolitaine de Montréal, 2011. *Un Grand Montréal attractif, compétitif et durable. Plan métropolitain d'aménagement et de développement* (PMAD) (online). http://pmad.ca/fileadmin/user_upload/pmad2011/documentation/20111208_pmad.pdf (Page visited on May 6, 2014)

Figure 1.14: Turcot Interchange (Photo Credit: Philippe Poullaouec-Gonidec, 2011)

trains traveling over the Autoroute 20 gateway corridor, as well as the vast industrial wasteland, make this corridor an important area for metropolitan planning and development. Consultations on revitalization issues concerning the old industrial sectors, the reuse of the Lachine Canal and its surroundings, the reconstruction of highway infrastructure, and the consolidation of public transportation equipment becomes more essential to reduce fragmentation of the territory.

Vision

Montreal's International Gateway Corridor

2

The different city entrance paths that were closely laid out throughout the development of the southwestern part of the Island of Montreal illustrate the splintered dynamics of the urban territories found in many cities around the world.[1] The territories are cut by the multiple linear infrastructures that pass through them, and that divide the residential neighborhoods, the industrial sectors and the natural wasteland into assorted pieces. In the Montreal context, it must be admitted that this territory is not only splintered physically, but also politically and administratively. In addition to crossing several distinct municipal boundaries, each infrastructure is managed and maintained by various administrative entities.

However, the absence of a shared platform for dialog becomes a major constraint to developing a collaborative planning process; all projects were previously developed in an isolated manner. The disagreement between the various governmental administrations on the Turcot complex reconstruction project deeply divided the main stakeholders responsible for the territorial planning, in relation to the future of this highway infrastructure project. In the wake of this debate, it is important to remember that the broader reflections about the development of the Montreal west entrance pathway were suggested both in a report of the Bureau d'audiences publiques sur l'environment (BAPE) on the redevelopment project of the Dorval interchange and in a government decree on the Turcot Interchange reconstruction project.[2]

It is in this particular context that the ministère des Transports du Québec (MTQ) gathered a group of professionals from several agencies concerned by this project to begin a collaborative planning process for Autoroute 20. Concerns regarding the city entrance development, still minimal at the time, nonetheless demonstrated a significant interest in this territory, and especially a more coherent interaction between the infrastructure and the adjacent areas. Illustrating the complexity on the governmental level, this committee was formed by about twenty federal, provincial, regional, municipal, local, and even private organizations linked to transportation infrastructures. The multiplicity of participants reflects the variety of interests and stakes for this territory.

---

1. Graham, S. and S. Marvin, 2001. *Splintering Urbanism, Networked Infrastructures, technological mobilities and the urban condition*. Routledge, London.
2. Gazette officielle du Québec, November 24, 2010; decree 890-2010.

Work table on the Autoroute 20 gateway corridor project.

MUNICIPAL LEVEL
- City of Dorval;
- City of Montreal;
    › Bureau du design;
    › Bureau du plan;
    › Division de l'urbanisme;
    › Borough of Côte-des-Neiges - Notre-Dame-de-Grâce;
    › Borough of Lachine;
    › Borough of Le Sud-Ouest;
    › Borough of Ville-Marie.
- City of Montreal West;
- City of Westmount.

REGIONAL LEVEL
- Conférence régionale des élus de Montréal (CRÉ);
- Communauté métropolitaine de Montréal (CMM).

PROVINCIAL LEVEL
- Ministère des Affaires municipales, des Régions et de l'Occupation du territoire (MAMROT);
- Ministère des Transports du Québec (MTQ), Direction of the Island of Montreal;
- Ministère du Tourisme.

PUBLIC AGENCIES OF TRANSPORTATION
- Aéroport de Montréal;
- Agence métropolitaine de transport (AMT).

RAILWAY NETWORK
- Canadian Pacific (CP);
- Canadian National (CN).

Heritage Conservation Organizations
- Parks Canada (Lachine Canal).

Considering this context of opportunities, the MTQ entrusted the Chair in landscape and environmental design of the University of Montreal to conduct a project of international scope aimed at providing urban planning visions for the Autoroute 20 gateway corridor between the Montréal-Trudeau Airport and downtown Montreal. The international ideas competition *YUL/MTL: Moving Landscapes* and the *Workshop_atelier/terrain (WAT) UNESCO Montreal 2011* were developed inside this mandate. Their objective was to share ideas on the future of the

Figure 2.1: The strategic vision of the city entrance corridor is based on simultaneously reading the issues, aspirations and opportunities of the latter.

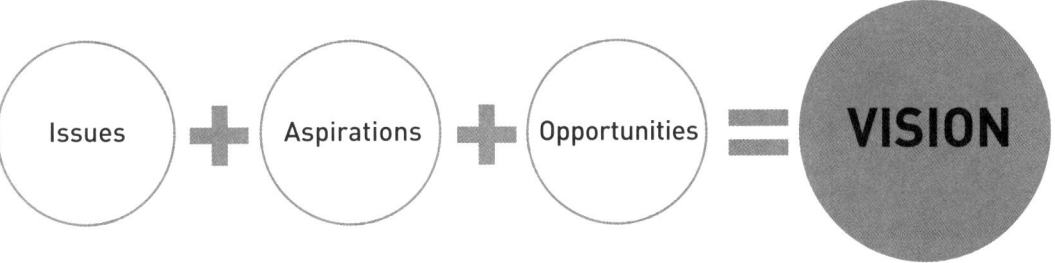

Autoroute 20 gateway corridor. Before these two ideation exercises, a planning process was necessary for all committee members collaborating on this territory to share their visions.

Given the urgency to act on a quite unstructured territory and the conflicting relationships between the parties involved in the debate, it became essential to reveal the shared values that can contribute to an acceptable decision for all.[3]

However, the visioning processes revealed several challenges; perhaps the most important one was to create a climate of confidence that promoted an open debate likely to exceed the individual positions of the stakeholders.

To overcome this difficulty, the approach encouraged participants to freely express their appreciation of the issues while paying attention to the viewpoints shared during discussions. To this end, individual interviews allowed them to create a general portrait of the issues, aspirations, and opportunities of the targeted territory as perceived by the stakeholders (Figure 2.1). This portrait allowed reading the stakeholders' concerns without only focusing on benefits, but also on actions that would make the enhancement and development of the territory possible.

---

3. Rauws, W et T. van Dijk, 2013. A design approach to forge visions that amplify paths of periurban development, *Environment and Planning B: Planning and Design*, 40: 254-270.

The challenge of the YUL/MTL project was to gather numerous territorial stakeholders around a common project to redefine the Autoroute 20 gateway corridor landscape.

Five questions, supplemented by an exercise of participatory mapping, were discussed with the members of the work table on the Autoroute 20 gateway corridor. They focused on the following themes:

- What is a city entrance?
- What is the identity of Montreal?
- What is a city entrance for Montreal?
- What are the special features of the Airport-Downtown entrance?
- What are your aspirations in regards to the Autoroute 20 corridor?

These questions first aimed to create a shared vision of the city entrance concept for all members of the work table. They then sought to compare, even contrast, the stakes of the real territory with the desired transformation of the latter into a city entrance corridor. This strategy also allowed each participant to take a necessary step back in regards to the daily planning challenges of the urban corridor.

Based on a thematic content analysis, the results of these discussions were communicated to all committee members while ensuring the responses were anonymous, with a graphic report in the shape of a radar. In doing so, it was possible for the participants to assess the main converging and diverging elements emerging from this portrait while having the opportunity to find their own viewpoints. In fact, these graphs consolidated, in their center, the recurring comments of several stakeholders who consequently had a chance to find the base of a consensus. More singular comments, just as important, are also found around the radar because they reflect a wider diversity of expressed viewpoints.

By opting for a simultaneous release of the results to all stakeholders, a quick consensus was found on the planning vision that guided the proposals' creation during the international ideas competition and the WAT_UNESCO Montreal 2011.

The lines that follow are intended to present (in summary form) the issues, aspirations, and opportunities of the Autoroute 20 gateway corridor that were identified during this crucial phase of the project.

Characterization of the Projects Issues, Aspirations and Opportunities

What is a city entrance?
The city entrance is inseparable from the identity of a city. It must carry this identity while expressing the personality and ambiance of a city and the meaning of the territory. It is an opportunity to reveal its values, culture, knowledge, and vitality (the heartbeat of a city).

In this sense, the entrance is linked to the image of the city, like a snapshot taken at a specific time. Moreover, the city entrance is the visitor's first contact. It is the first image of a city, the first impression.

The city entrance is also a showcase of everyday life, of the urban life—a showcase that announces what is coming.

The city entrance is not a gate, but a path. It is a dynamic concept related to the experience and movement, the progress and the landscapes that are seen successively. This progress can be seen through the intensification of the city up to the discovery of its heart, its climax. Thresholds that reflect breaking points in the landscape or surprises and discovery also mark the city entrance. In this sense, the city entrance is strongly linked to the user's experience.

Sharing issues and aspirations of the YUL/MTL corridor helped define a strategic vision that targets the enhancement as much as the development of highway and territorial landscapes.

Figure 2.2: Graph of distribution of the answers to the question: According to you, what is a city entrance?

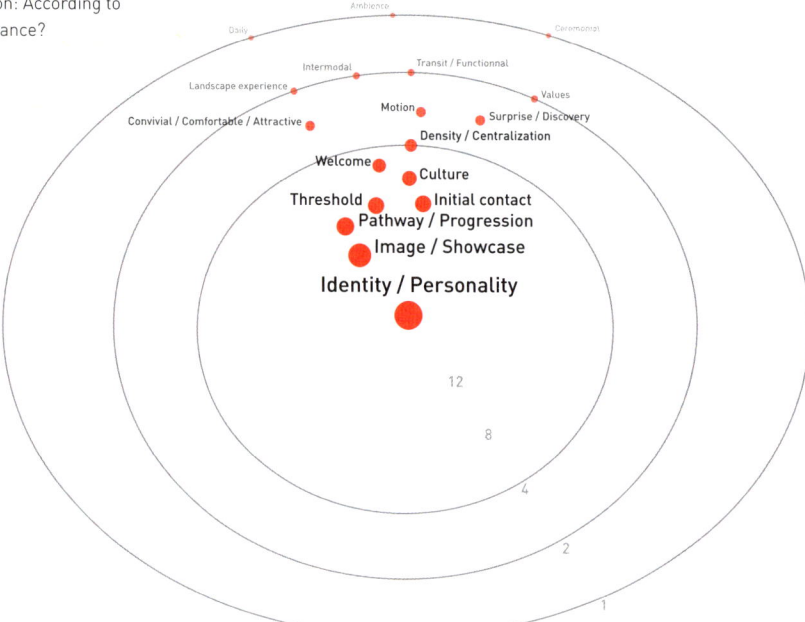

Figure 2.3: Graph of distribution of the answers to the question: What does the image or identity of Montreal call on?

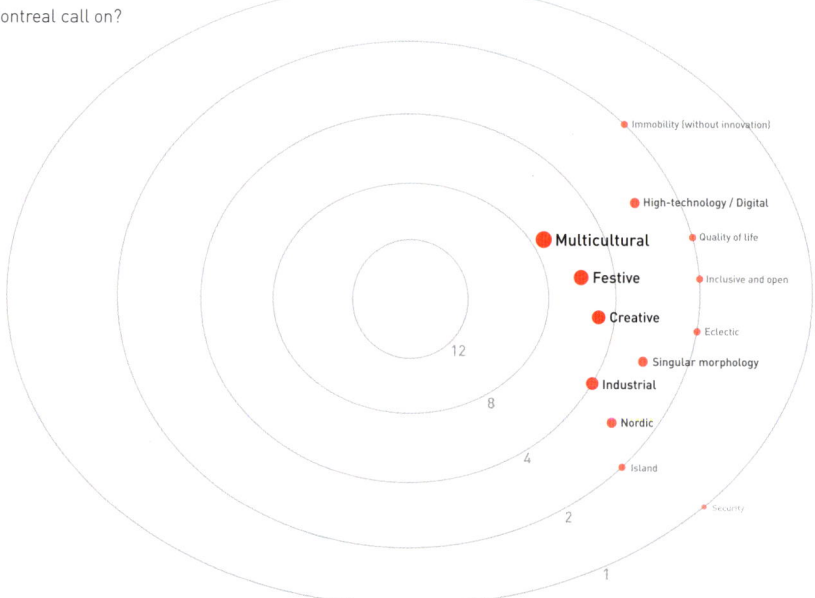

YUL/MTL   VISION

Figure 2.4: Graph of distribution of the answers to the question: What is a city entrance for Montreal?

Figure 2.5: Graph of distribution of the answers to the question: What are the special features of the Airport-Downtown entrance?

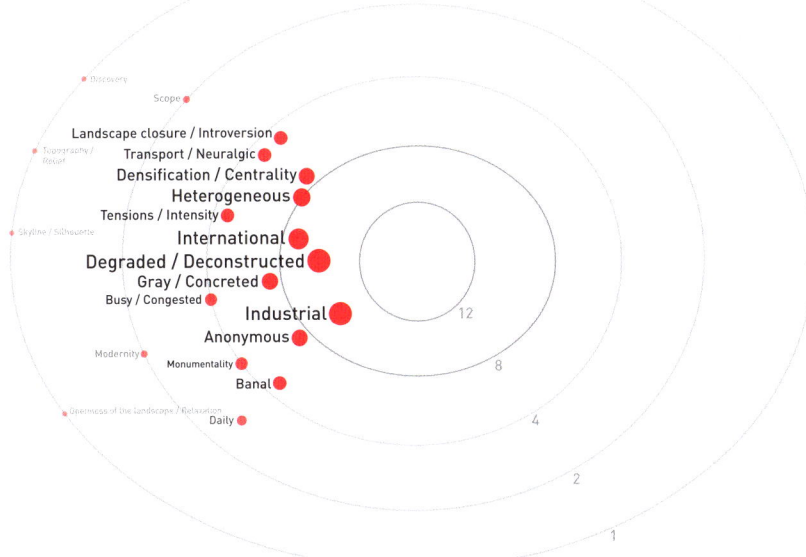

The city entrance is also associated with a functional notion of transit, a certain accumulation of transportation infrastructures that are often intermodal (railway, highway, public transportation). It is then immersed in the daily transportation experience for the urban periphery commuters (Figure 2.2).

What does the image or identity of Montreal call on?
Montreal identity is expressed through culture and creativity. The nomination of Montreal as a UNESCO City of Design encourages the creativity of the designers and emerging professionals who participate in imagining the city. In parallel, a multimedia industry emerged a few years ago, attracting a significant number of young people who live and work in Montreal. This new creative economy in formation is part of the identity of Montreal. Montreal also emerges as a city of festivals, pleasure and performing arts.

The industrial character of Montreal, still very present (ex.: port, working class neighborhoods and industrial areas) shows its urban heritage. In this respect, the Lachine Canal, which crosses the island in its southwestern part, is recognized as being the cradle of industrialization in Canada.

Montreal identity is also expressed through its multicultural character and its openness. This mix of cultures is also present in the landscape expressions of Montreal, which are manifested in their eclectic and heterogeneous forms. This underlines the sociocultural and landscape differentiation of its neighborhoods since each has its own identity. Despite these differences, or even because of these differences, it is good to live in this city. A safe city offers a nice quality of life.

Finally, Montreal is a Nordic city, marked by the rhythm of the seasons that creates different landscape moods and experiences throughout the year (Figure 2.3).

What is a city entrance for Montreal?
Montreal is an island; several of its city entrances are crossed on bridges. This island condition shows the important role of watercourses (ex.: St. Lawrence River) in the landscape perception of city entrances. The passage of water creates a clear division between the city and its periphery. Whereas elsewhere, travelers often cross-large peri-urban regions without definite borders before arriving downtown, in Montreal, the arrival point is clear, especially in its connections to the south.

These watercourses also allow an opening of the city. Thus, when crossing bridges, drivers benefit from a few of the most emblematic viewpoints on Montreal: the downtown skyline, Mount Royal ,and the river. This facilitates the legibility of the city and its progression toward a centrality. This feature highlights the singularity of Montreal and the diversity of its built environment.

Nevertheless, outside the large crossings of the St. Lawrence River, the city entrances of Montreal are more often referred to as ordinary and functional. On its northern flank, the narrower watercourses allow more discrete crossings where the landscape opens almost only to the road (Figure 2.4).

What are the special features of the Airport-Downtown entrance?
This highway entrance corridor is overall degraded and unstructured. Signs of infrastructure aging and the presence of numerous brownfield sites contribute to a negative perception of this corridor. The industrial character of this part of Montreal seems obsolete due to the physical condition of the buildings and infrastructures that punctuate the territory. From the highway, we can see a heterogeneous urban environment composed of industrial backyards, vacant spaces, or warehouses.

However, industrial remains testify to the existence of a golden past. As the cradle of industrialization in Canada, the Lachine Canal is an important place in the national economic development history . Its industrial artifacts have great symbolic significance.

This corridor is also a central pillar of the national transport system. Traffic is dense and the corridor is usually very busy, even regularly congested. The predominance of transportation infrastructures, of concrete and shades of gray, promotes a degree of anonymity, a certain banality. Vegetation can mainly be found in the background. In regards to user's experience, the path views offer both openings on adjacent territories and visual closing on the road. A sense of introversion, suffocation, and tension is occasionally generated.

Despite the banality of places, the international scope of the Autoroute 20 gateway corridor is emphasized, since it constitutes the main link between downtown and Montreal-Trudeau airport. It is also an entrance used daily by commuters to access downtown.

Its modern character distinguishes this city entrance corridor. Several of its major civil engineering masterpieces date back from the 1960s. They testify to a period during which Montreal was still the metropolis of Canada and received the attention of the whole world, in particular for the 1967 Universal Exhibition. The monumental aspect of the infrastructures and, in particular, of the Turcot Interchange, thus generated a certain nostalgic pride (Figure 2.5).

Aspirations
The clear contrast between, understanding the city entrance concept as a reflection of the city identity and the unstructured nature of the Autoroute 20 corridor brings up many aspirations for the future of this territory. In order to simplify the reading, all aspirations were gathered and grouped around the following themes: identity, sustainable planning, and pathway.

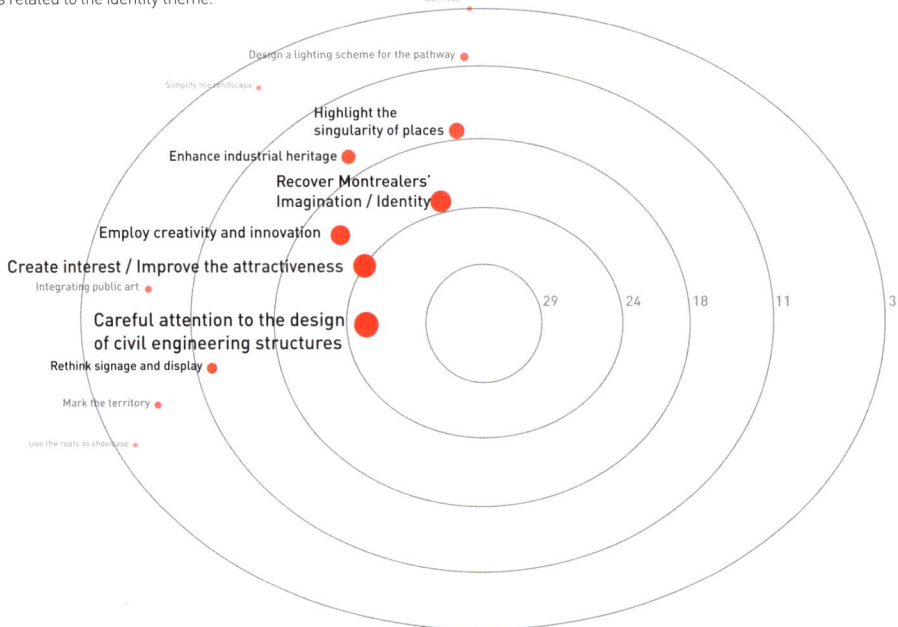

Figure 2.6: Distribution chart of the aspirations related to the identity theme.

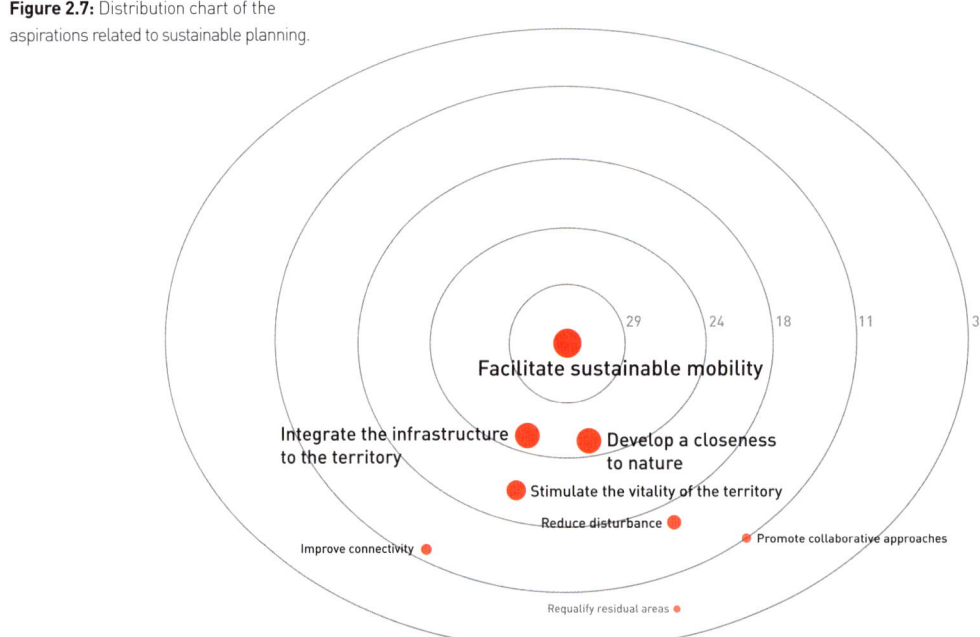

Figure 2.7: Distribution chart of the aspirations related to sustainable planning.

YUL/MTL   VISION

Figure 2.8: Distribution chart of the aspirations related to the pathway.

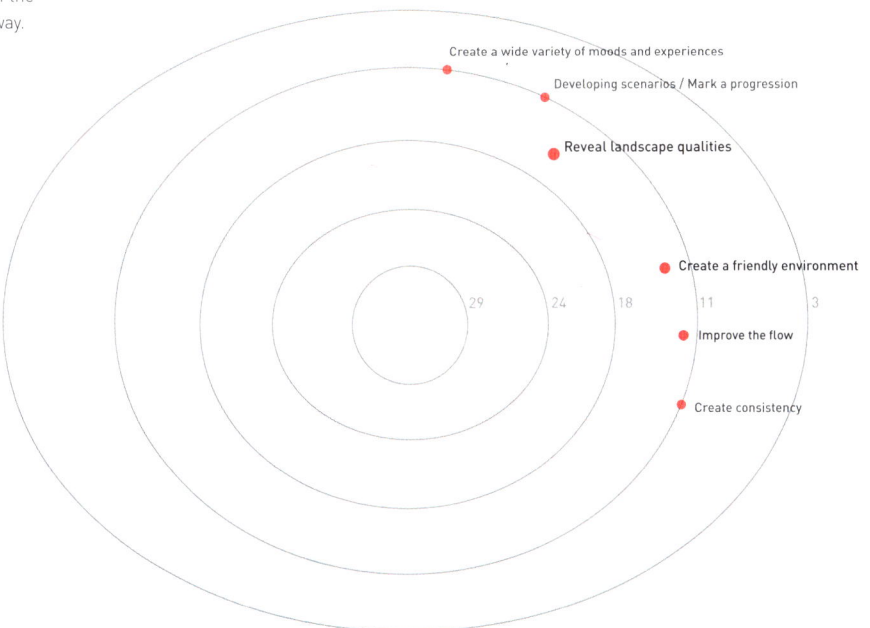

Identity
In regards to identity, it is desired that the city entrance corridor becomes more attractive, with particular attention to the design of civil engineering structures. This aspiration evokes the potential of infrastructure to become emblematic for a city, without denying the singular character of the place. Thus, several suggested that the industrial heritage be revealed as well as the neighborhoods' landscape peculiarities. There is consensus desire to enhance the identity of Montreal, whether or not it is already recognized, in the making, or still needs to be created. In this sense, the project appears as an attractive opportunity to make Montreal live and radiate as the UNESCO City of Design.

Some comments favor simplifying and facilitating the legibility of the city entrance corridor by improving signage, especially taking into account the speed with which the highway drivers experience it (Figure 2.6).

Sustainable planning
The comments also expressed using the sustainable planning concept into a specific manner and in accordance with local realities of the

Figure 2.9: Diagram of the limits of the city entrance territory drawn by local experts.

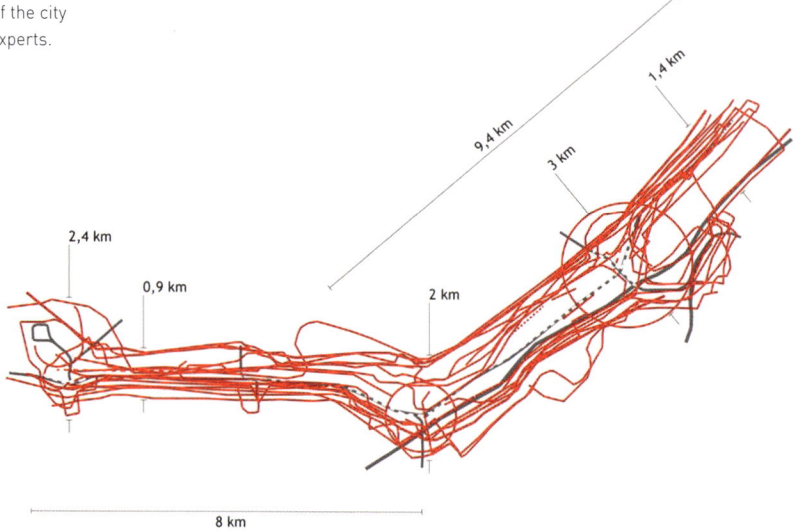

targeted area. Thus, the concepts of sustainable mobility, nature, and vitality play an important role in the future of the Autoroute 20 corridor. The dynamics of the public transportation infrastructure, the corridor's assets, the enhancement of natural spaces, and the improvement of brownfield sites represent an action plan toward renewing the highway corridor.

Beyond these specific elements, a coherent, transversal, and integrated approach of the territory project is expected through the sustainable planning concept (Figure 2.7).

Pathway
The comments concerning the notion of pathway refer to the idea of extending the reach of a city entrance in order to explore the collaborative planning of a pathway that covers several kilometers. Thus, the aspirations related to this concept are targeting the experience of travelling through the corridor. Therefore, the current reflection aims to contribute to scripting this crossing to schedule, mark, and structure the pathway in a consistent, fluid, and user-friendly manner (Figure 2.8).

Limits of territory and sites of opportunities
During this exercise, each respondent was invited, at the end of the talk, to draw an image of the studied area, the limits of the territory considered in the project. It clearly identified the boundaries of the "playing field" offered to designers during the subsequent phase of ideation. The attached figure then presents a superposition of all the collected drawings. Thus, Figure 2.9 helped identify the contours of a territory that could cover a width of one to three kilometers in some

Figure 2.10: Diagram of the sites of opportunities of the city entrance territory drawn by experts.

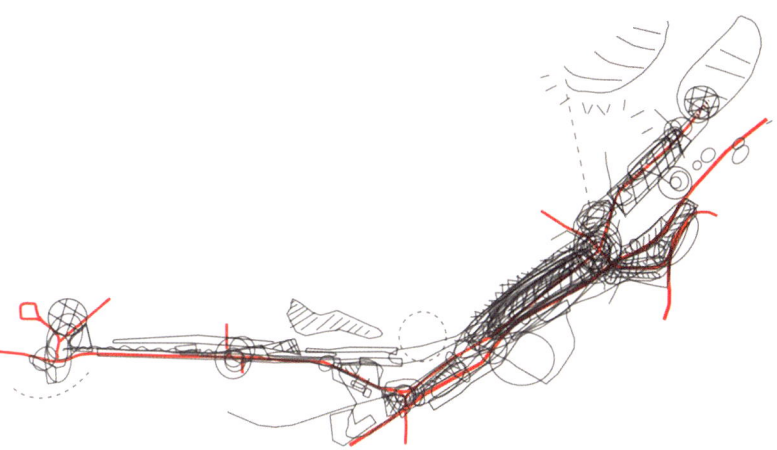

places, thus integrating the urbanized environments adjacent to the highway. The resulting perimeter allowed understanding better that the city entrance is more about a territory project, and not only about a highway project.

The mapping exercise also revealed the existence of more than fifty project opportunity sites. These sites are found as much on the highway right-of-way as in the adjacent neighborhoods, and involve many public and private stakeholders. While some of the opportunities sites are in the masterplanning process or under construction, others are waiting for redevelopment in the near future (Figure 2.10).

The determination of the intervention perimeter and project opportunity sites strengthened the collective recognition of issues regarding the city entrance corridor. In addition, all stakeholders were able to note, perhaps for the first time, the immense renewal potential of the city entrance corridor and the need for collaborative actions because of the imminence of major projects. In the end, this exercise will have made possible a better understanding of the territory, as it exposed a mutual learning of the issues, aspirations, and opportunities related to the city entrance corridor.

Figure 2.11: Map summarizing the perimeter of intervention and places of opportunities for the Autoroute 20 gateway corridor project.

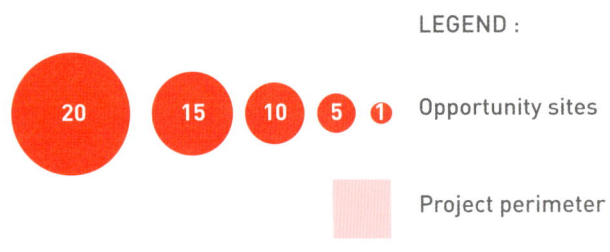

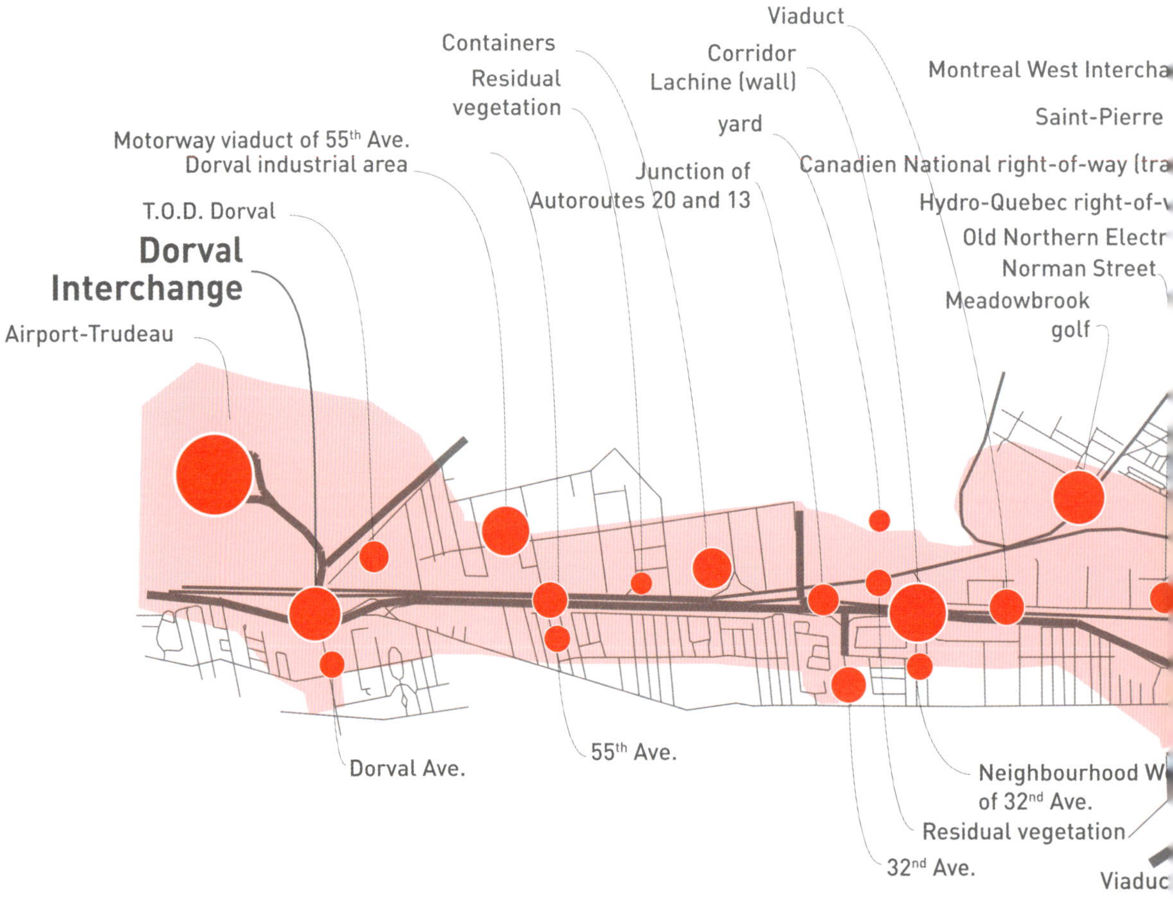

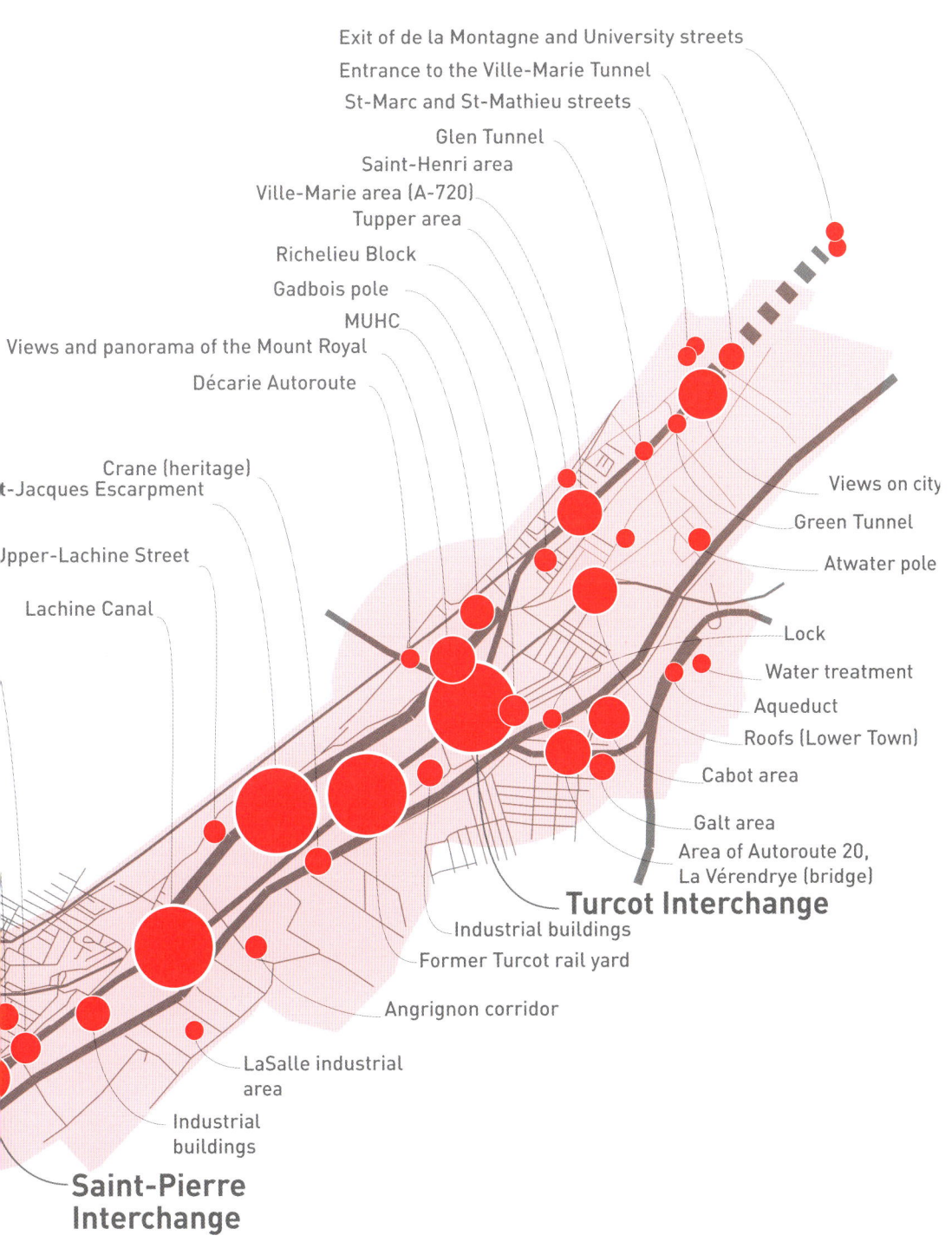

Vision Statement

This vision statement is an excerpt from the Rules and Regulations document of the *YUL/MTL: Moving Landscapes* international ideas competition released in June 2011. It has been prepared on the basis of the synthesis of interviews with members of the working table on the Autoroute 20 gateway corridor and approved unanimously by them prior to the launch of the competition.

Rather than a simple transportation corridor, Montreal's international gateway reveals in its functional asset the city's identity. The competition thus launches a genuine territory project that requires planning at multiple scales:

- At the regional scale, the intent is to create a coherent strategy for the entire corridor that takes into account all the possible experiences of the local landscape;

- At the local scale, the proposals should pay close attention to the quality of the interactions between living environments, the highway and other transportation infrastructures.

The proposed gateway corridor framework should simultaneously explore the programmatic complexity of its urban realm composed of:

- Transportation infrastructures that generate entry and exit routes to and from the downtown area: In the particular case of Montreal's international gateway, these infrastructures are not limited to the highway, but also include two railway lines, as well as a recreational boating canal and a bicycle path. These different infrastructures provide diverse experiences from the vistas they offer as well as for the traffic flow they allow.
- Living and working environments: Entering the city, one discovers living environments, which gradually densify from sprawling suburbs to older neighbourhoods. It is also an initiation to one of Montreal's oldest industrial landscapes, which once was the industrial cradle of Canada, but whose vitality has strongly decreased and is in need of revitalization.
- Natural environments: Although they are sometimes designated as parks and conservation areas, like de Lachine Canal National Historic Site, the main natural spaces along this corridor are mostly residual biomasses which necessitate enhancement such as is the case of the Saint-Jacques Escarpment.

The collaborative process with the public and private stakeholders concerned with the gateway corridor has helped to pinpoint three main themes which should guide the contestants in their reflections on the future of this territory.

1st Theme: An evolving and emblematic landscape project for the metropolitan area

The competition focusses above all on the issue of landscapes which are perceived through the general appreciation of emblematic vistas (ex.: Mount-Royal or the downtown skyline) as well as through the recognition of the multiple qualities of the surrounding living environments. Thus, multiple views of the territory are revealed which carry social values and aspirations such as:

- Aesthetic and visual values: these are revealed by the perceptual features of the environment, by the visual appeal of its scales, vistas and deployed horizons (from foreground to background), and by the generated kinetic effects;
- Environmental values: these are revealed simultaneously when taking into account both the natural environment and the environmental innovation (ex.: green infrastructure) of the highway project;
- Economical, social, and cultural (heritage and innovation) values: these values mainly stem from the industrial character of the gateway corridor and from the need to highlight this character within Montreal's multiculturalism as well as in the cultural creativity it has to offer;
- Identity values: they are expressed by the enhancement of emblematic landmarks but also according to the changing seasons (ex. nordic condition).

Subject to the ever-changing, dynamic, and never time-bound values, the landscape project therefore calls for an open attitude towards enhancement and development opportunities as well as for the creation of new landscapes.

2nd Theme: A scenographic composition of the corridor experiences

In the particular context of Montreal's international gateway corridor, the present reflection focuses on the user's point-of-view, whose perception and experience of the landscapes is modulated by speed. This movement-based experience is perceived through the blurred, rapid-moving foreground, the slow-moving middle distance, and the seemingly motionless background.

The landscape is additionally expressed by the gradual intensification of the urban fabric. However, this progression, far from being linear and regular, is punctuated with nodes and landmarks such as the network junctions and emblematic vistas. The need for a scenographic composition of these landscape elements should increase the legibility and coherence of the gateway corridor's structure by taking into account:

- The driver and passenger positions as well as the variation due to different transportation modes (car, bus, passenger trains, etc.);
- The experience of entering and exiting the city;
- The ambiances related to day and night passages as well as those related to seasonal changes (spring, summer, fall, winter).

The experiences of residents living close the autoroute must also be taken into account. While benefitting its access, the residents also endure the frenetic ingoing and outgoing traffic of a major Canadian city.

3rd Theme: A collaborative approach for sustainable urban development

The gathering of the issues and aspirations expressed by the public and private stakeholders consulted has allowed for the creation of the structure of a specific collaborative approach for the sustainable development of Montreal's international gateway corridor. This approach emphasized the perceptual voids projected by the Autoroute 20's image. Thus, the sustainable development of the gateway corridor primarily aims to stimulate the territory's vitality.

The competition area has in this regard many undeniable assets. It is an infrastructure corridor composed of multiple transportation modes. This area is also close to living environments as well as rich ecological areas such as the Saint-Jacques Escarpment. Nevertheless, the gateway corridor has many underused gaps and brownfields in need of revitalization. The territorial project should illustrate how transportation projects can heighten the vitality of these areas and how in turn they can frame the infrastructure's landscapes.

Ultimately, Montreal's international gateway corridor is a territorial project which aims to showcase the city's vitality and creativity.

Conception guidelines

Guidelines for an emblematic landscape project for the city

- Creating a strong and expressive statement:
    - Expressing Montreal's distinct image of technological innovation, creativity, design, and performing arts;
    - Taking into account the different seasonal conditions of a nordic city.

- Highlighting the uniqueness of Montreal's setting:
    › Enhancing its industrial and built heritage;
    › Revealing the townscape qualities of local areas along the gateway corridor.
- Demonstrating creativity and innovation:
    › Improving the urban design and architectural features of the infrastructures (street furniture, equipment, structures, etc.);
    › Creating varied ambiances that can be modulated by lighting effects or by temporary installations.

Guidelines for the scenographic composition of the corridor's experiences

- Planning the gateway corridor as a 21st century urban promenade:
    › Marking the progression towards downtown or inversely towards the airport;
    › Taking into account the different gateway experiences (highway, railway, canal);
    › Envision new interfaces that are coherent with the concept of an urban gateway while enhancing the quality of living environments.
- Insuring the coherence of the interventions:
    › Improving the legibility of the urban environment by redesigning the signage, the advertising structures, and the furniture;
    › Creating a safe and user-friendly environment.

Guidelines for a collaborative approach to sustainable urban development

- Stimulating the vitality of adjacent neighbourhoods:
    › Revitalizing residual spaces;
    › Structuring the boundaries and interconnections between the infrastructure and the adjacent areas (ex.: landscape, living and built environments).
- Increasing the proximity to nature:
    › Creating greener infrastructures (ex.: noise barriers) and adjacent areas (ex.: roof) so they may become an expressive feature of the environmental urban design avant-garde;
    › Networking natural and recreational areas.
- Integrating the infrastructures to the territory and vice versa
    › Improving connections between neighbouring areas;
    › Reducing nuisances (ex.: quality of life – air, visual and sound pollution).

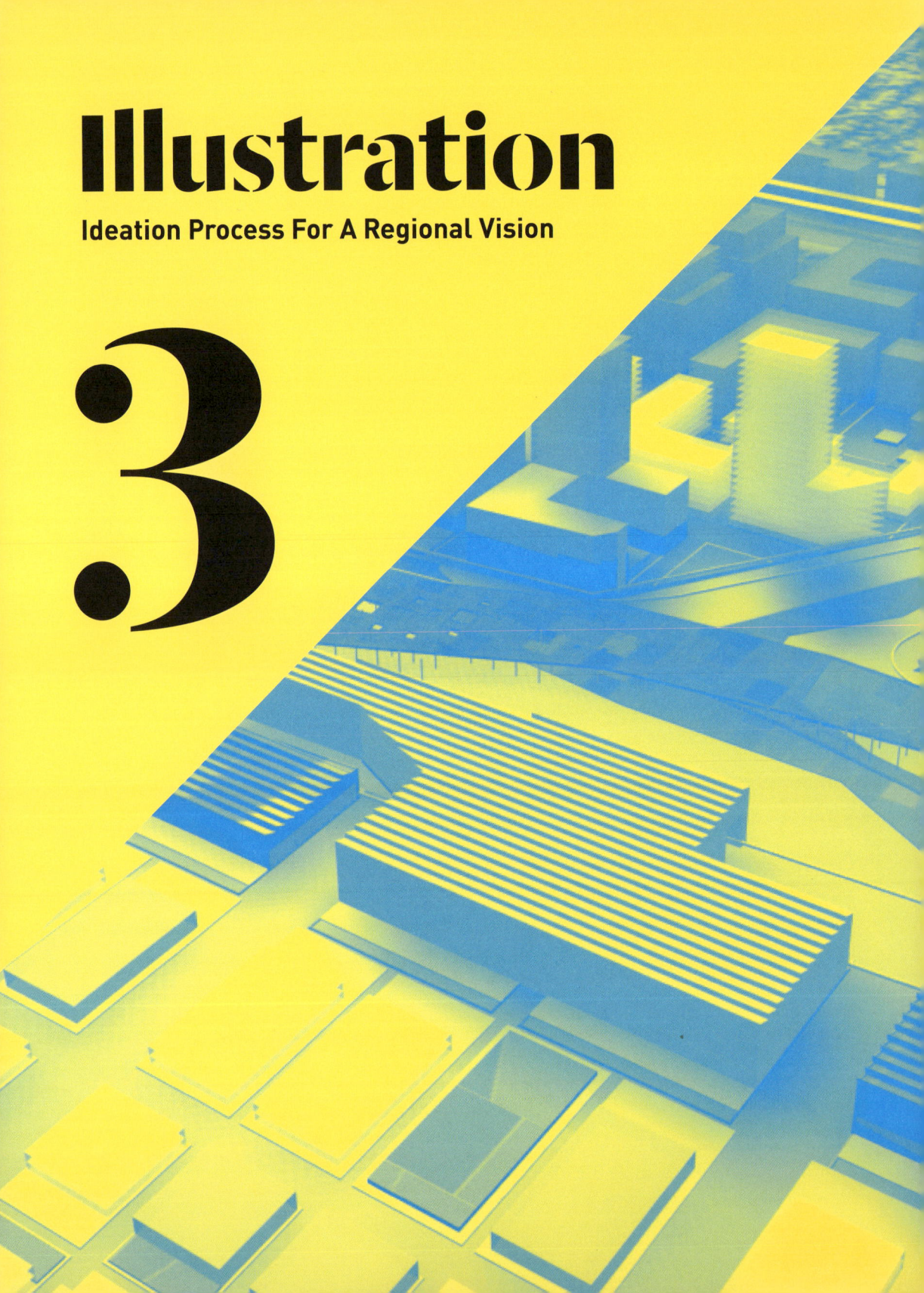

Illustration
Ideation Process For A Regional Vision

3

An integral, even central, part of the *YUL/MTL: Moving Landscapes* project, the ideation phase aimed to bring multidisciplinary knowledge and dialog to the involved experts, to generate flows of plural knowledge. The direct and involved contributions of professionals and students from different disciplines of design and planning (architecture, landscape architecture, urban design, and urban planning) helped to improve the territorial vision developed by the work table of the Autoroute 20 gateway corridor, as well as create new viewpoints in urban design. It is important to recall that the ideation approach used in the highway redefinition process was a tool of collaborative and territorial planning. It helped illustrate the territorial vision expressed by all of the involved territorial stakeholders. In that respect, this territorial vision has been the bedrock of this ideation exercise.

In order to consider the site as a complex whole, two separate processes of ideation[1] were implemented, the international *YUL/MTL* ideas competition and a design workshop, entitled *WAT_UNESCO Montreal 2011*. Complementary to the other, the two ideation exercises each had their own purposes and modes of operation.

Geared toward professionals, the international ideas competition[2] targeted the development of broad planning strategies that concerned the whole territory of the gateway corridor, representing seventeen kilometers of highway and its adjacent neighborhoods. In contrast, WAT_UNESCO[3] was primarily an international activity for students and professors, similar to an "ideation workshop in urban design", for which the corridor was divided into six segments. This difference in scale between the two ideation activities gave the possibility to place

---

1. The experimentation on the use of various ideation models in design is justified in particular in regards to a governmental, municipal and provincial will to promote new processes to better build the city, which was affirmed especially by the publication of *Cahiers des bonnes pratiques en design* (see Lemieux, 2008), as well as in the context of the designation *Montreal, UNESCO City of design*.
2. The ideas competition is a way to call for proposals on architectural, urban and landscape issues. Not resulting in ordering the project, the ideas competition is usually an open process favoring the exploration and emergence of innovative concepts.
3. The Workshop_atelier/terrain (WAT_UNESCO) are organized every year since 2003 by the CUPEUM in different regions of the world (Europe, North Africa, Middle East and Asia) and bring together an international network of more than 20 academic institutions. The purpose of this ideation activity is to provide answers to the problems of urban development that call out to major social, cultural and environmental issues.

The interest generated by the ideas competition demonstrates the international scope of the highway redefinition themes and the quality of city entrances.

emphasis on separate elements of the corridor, which are the most significant territorial challenges.

With its territorial ambition, the results of the ideas competition insist on the dynamics of transportation networks and the constraints or potential offered by these networks for the development of living environments adjacent to highway infrastructure. Thus, several proposals submitted for the ideas competition suggest restructuring the highway network, adding lines of public transport or bicycle lanes to connect the whole territory. Targeting smaller areas, WAT_UNESCO provided the possibility to explore interactions between neighborhoods adjacent to the infrastructure and the infrastructures themselves, as well as the quality of life in these neighborhoods.

In addition to a complementary scale, the two ideation activities were programmed to succeed in a timely manner, to ensure a close link between the two times of reflection (and idea production). Thus, WAT_UNESCO was simultaneously launched with the announcement of the winners of the international ideas competition. Present in Montreal for this announcement, the winners were able to show and explain to students and faculty participants in WAT_UNESCO the basic principles of their planning strategy and intervention. Thus, the strategies set forth by the winners of the international ideas competition directly inspired and guided the design work of the students and professors who could apply, test, and operationalize, for reduced areas of the corridor, the overall strategies developed by the professionals. The time overlap of the two activities provided the process with consistency between the developed principles.

The ideas competition and the design workshop nevertheless differed on the relationship between the designers and the local

The designers' proposals allow renewing the viewpoint that is currently taken on the territory.

experts in charge of developing the territory. Thus, the international ideas competition allowed creating a voluntary distance between the designers' work (professionals) and that of local experts, explaining the importance of having committed to a collaborative approach ahead of the competition so that designers could obtain maximum information to develop relevant proposals and respond to local aspirations. In fact, based on this vision, the contribution of the designers was regarded as an independent and outside viewpoint to generate constructive perspectives and criticisms in relation to the local, targeted issues.

As for the design workshop (WAT_UNESCO), it helped maintain a closer relationship between the participants (students and professors) and local experts, since periods of exchange were planned. Gathered in the same workspace, participants of this international activity generated a climate of greater collegiality.

The purpose of this chapter is to present the main results of both the international ideas competition and the WAT_UNESCO. It brings together the illustrations and outcomes of the three proposals from the winners of the international ideas competition. These elements are introduced by the general remarks of the jury, who underlined the innovative effort of the competition itself and its true relevance to the necessity for a collaborative approach to highway corridor landscape planning. Subsequently, the twelve proposals designed by the participants (students supervised by professors) of WAT_UNESCO are presented. They were grouped by area of intervention, that is to say, areas identified by the members of the work table on Autoroute 20 gateway corridor. A description of the major issues targeted by this committee for each of the areas precedes the presentation of proposals.

YUL/MTL international ideas competition

Open to design professionals, the *YUL/MTL: Moving Landscapes* international ideas competition[4] generated sixty-one proposals from twenty-two countries, thus demonstrating the resonance of the theme of city entrances and highway redefinition in several regions of the world.

The proposals were assessed over a period of two days at the end of October 2011. The jury was composed of various professionals, with expertise in transportation, engineering, architecture, urban design, landscape, and scenography.

Upon reading the submitted proposals, the planning goals and challenges raised by the international ideas competition revealed multiple ideas. Through its deliberations, the members of the jury looked for proposals that integrated all of the challenges and concerns of planning to formulate an intervention strategy that defines a real territorial project. In this sense, the proposals that raised the most interest were those that allowed inducing a simultaneous reflection on the highway, other transportation infrastructures, and the articulation of adjacent living environments and ecological systems.

The jury thus stressed the importance of working simultaneously on the infrastructure and the territory. It falls directly within the purpose of the initial program of the competition and the ambitions formulated by the members of the committee of the Autoroute 20 corridor, that is to say, the urban planning vision established ahead of the competition. The jury noticed that collaboration between all stakeholders within a planning process was rare in infrastructure projects, and that the collaboration undertaken ahead of the *YUL/MTL* international ideas competition belonged to an innovative perspective, which renewed the planning methods that had to be put in place in the road and highway infrastructure redefinition projects.

One of the main findings issued by the members of the jury after reviewing the proposal packages is the enormous transformation potential of the Autoroute 20 gateway corridor territory. With respect to a currently abandoned corridor, the images drawn by the designers envision a lively, urban, dense territory, and consequently change the perception that one can have of the present environment. The

---

4. For more information about this competition, see: (http://mtlunescodesign.com/fr/projet/Concours-international-didees-YUL/MTL-Paysages-en-mouvement)

proposals generated during this international ideas competition identify unsuspected potential. They also illustrate a range of territorial interventions, from simple to more complicated.

In order to guide the continuation of the collaborative planning process, the jury wished to exclude the rationale that a single proposal be put forward through awarding a first prize. The decision of the jury to assign three tied prizes emphasizes the complementarity of award-winning urban planning visions. Beyond bringing tangible and viable projects, the three winning proposals offer, in a complementary manner, a set of strategic planning concepts (macro-design) that consider the design of new identity-marking elements for the city entrance pathway as much as the redefinition of the relationship between transportation infrastructure and its adjacent living areas.

Far from contradicting or opposing themselves, the three winning visions complement each other. Some of the ideas laid out in each of them may overlap, or be juxtaposed. To this extent, they offer, according to the terms of the jury, a real "atlas of possibilities", which can serve as inspiration as much for Montreal as for other cities in which the insertion and redefinition of highway transportation infrastructures is a major challenge. In addition, it is important to mention that several other proposals of the international competition received recognition from the jury to complete this "atlas of possibilities", stressing the importance of retaining a palette of planning ideas. In fact, through its inclusive attitude of various viewpoints, the jury grasped very well the meaning and the role of the ideas competition in the planning process initiated in the *YUL/MTL: Moving Landscapes* project, which is to open the project's perspective and to illustrate the design guidelines.

The set of proposals constitutes an inspiring universe of possibilities for the development of highway transportation infrastructures.

The jury of the YUL/MTL international ideas competition was comprised of the following members:

Édouard François, architect and urban planner, France (president of the jury);
Pierre Bélanger, associate professor in landscape architecture, Harvard University, Graduate School of Design, USA;
Ken Greenberg, architect and urban designer, Greenberg Consultants Inc., Canada;
Florence Junca-Adenot, founder of the Urba 2015 Forum, Université du Québec à Montréal, Canada;
Anick La Bissonnière, architect and scenographer, Atelier Labi, Canada;
Jacques Verville, engineer, representative of the ministère des Transports du Québec, Canada.

Winning Projects

Undercover Montreal

Brown and Storey Architects, Toronto, Canada
James Brown, architect and urban designer, OAA
Kim Storey, architect and urban designer, OAA
Stephen King, B.Arch
Richard Averill, B.Arch
Matthew Unternahrer, trainee
John Duchene, B. L. A., OALA
Emma Brown, editor

Figure 3.1: New territorial topography of infrastructure that goes along and over Autoroute 20 in an asymmetrical manner. (Excerpt from the proposal of Brown and Storey Architects Inc., Canada)

Text submitted by Brown and Storey

Drawing on our research into the buried Garrison Creek Ravine – a major urban ravine buried in the nineteenth and twentieth centuries in the west end of Toronto – we have developed a new infrastructural instrument that takes its characteristics from an ecological source. This ravine-like isotropic figure creates an innovative approach to a process of regeneration of industrial wastelands, the 'collateral damage' caused by the construction of road infrastructure in the 1950s and 60s.

Within the seventeen-kilometer gateway corridor linking Montreal-Trudeau Airport to the downtown area, existing infrastructure has cut neighborhoods into parts, severed with circumstantial roads and dead ends, made improper conditions of backs and fronts, and has generally developed without quality or protocols. We propose a new network which we have named Undercover Montreal, which offers an integrated and bundled infrastructure – of novel ecologies of water storage, fast and slow infrastructural movements working together, and new linkages for the next generation of city building in a more plural environment. Not simply a subsystem, Undercover Montreal is an intentional and opportunistic network marking contact between communities.

The CCA publication, 'The Singularities of a Metropolitan Archetype', shows a series of figures prepared in 1966 by the City of Montreal Planning Department modeling hypothetical plans for 10 million inhabitants: the star-shaped metropolis, the galactic metropolis, and the concentrated metropolis. The concentrated metropolis prevailed. Meanwhile, the international corridor from airport to downtown is 'other': a vast residual territory of unrealized land value.

With Undercover Montreal we explore a model that more resembles the first of these figures – the star-shape – to make a new topography of infrastructure beside the hovering Autoroute 20. This new territory has been carved through the international gateway corridor lands to create a bundling of inhabitable, infrastructural entities. These co-evolve, adjust, and transform– taking the form of public parks, cyclist and pedestrian throughways, water storage, and flow channels, including the reinvigorated Lachine Canal, and public transport, including subways and raised rapid transit.

This all converges to make a seventeen-kilometer, continuous, diverse edge from the Montreal / Trudeau Airport to the tunnel entrances into downtown Montreal. The fingers of Undercover Montreal are a fine-grained, complex topography moving through an urban territory. This 'cover network' is placed, at one level, adjacent to existing neighborhood areas and meshing with lost ravines, railway spurs, and trails. At another level, these routes extend outward across

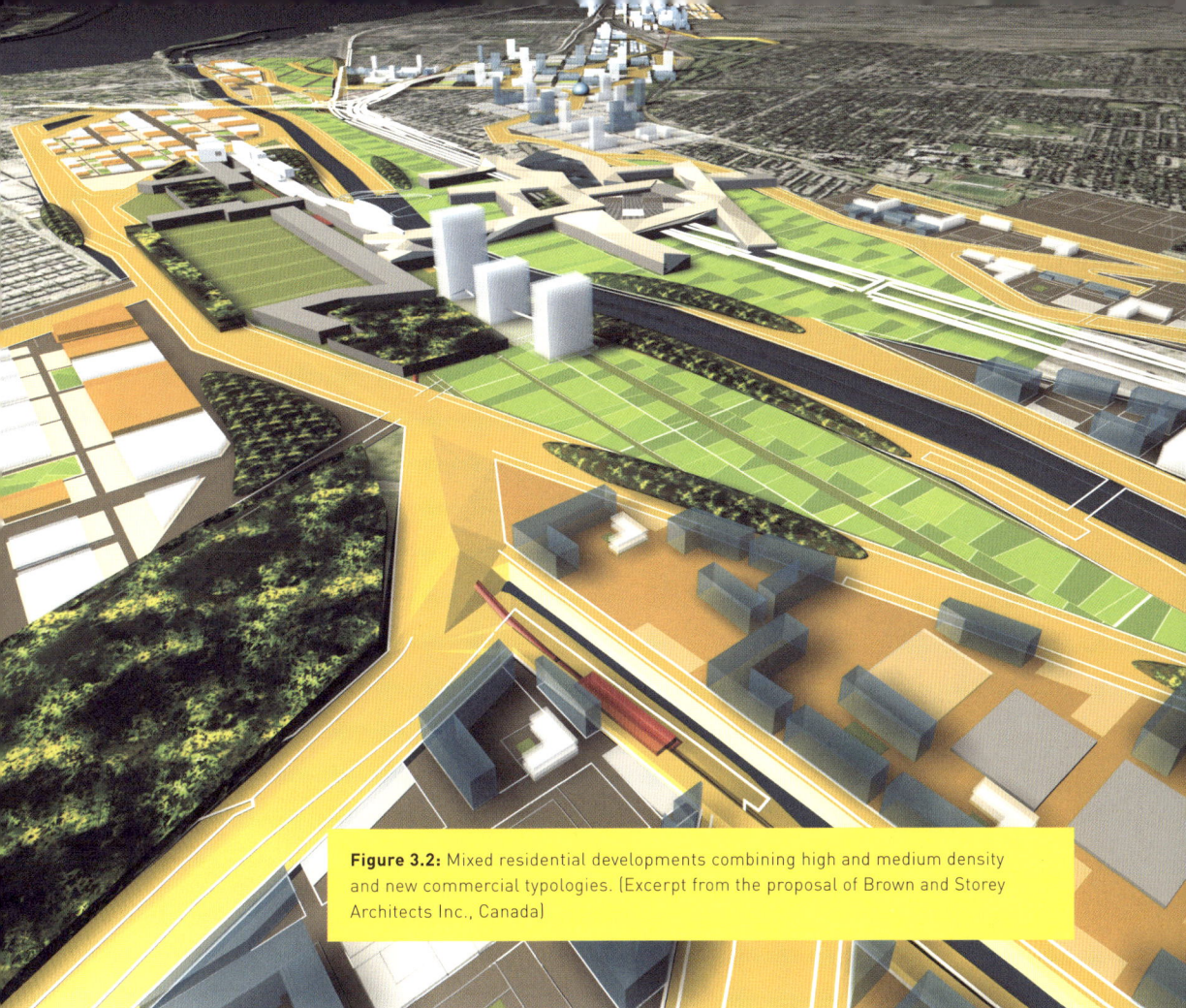

Figure 3.2: Mixed residential developments combining high and medium density and new commercial typologies. (Excerpt from the proposal of Brown and Storey Architects Inc., Canada)

the larger metropolitan territory as an organizer of new, large assemblages of selected public open spaces, as well as residential, industrial, and intensified commercial areas.

Undercover Montreal, carving and exploring the present territory, is a new isotropic form. Along its length it submerges and rises or meets the level of the surrounding landscape, connecting adjacent public spaces and inserting a new subway system and neighborhood precincts. Its fingers – 'jet streams' – make new spatial types, defined as 'strings' and 'large chunks'. They are provocative instigators for growth that are closely tied to the new infrastructure. This is a dynamic process that is self-sustaining and self-stimulating.

Figure 3.3: The star-shaped metropolis, carved out of Autoroute 20 gateway corridor, groups residential entities and infrastructures that are evolving, adjusting, and transforming simultaneously – public parks, cycling and pedestrian lanes, underground and surface public transportation. (Excerpt from the proposal of Brown and Storey Architects Inc., Canada)

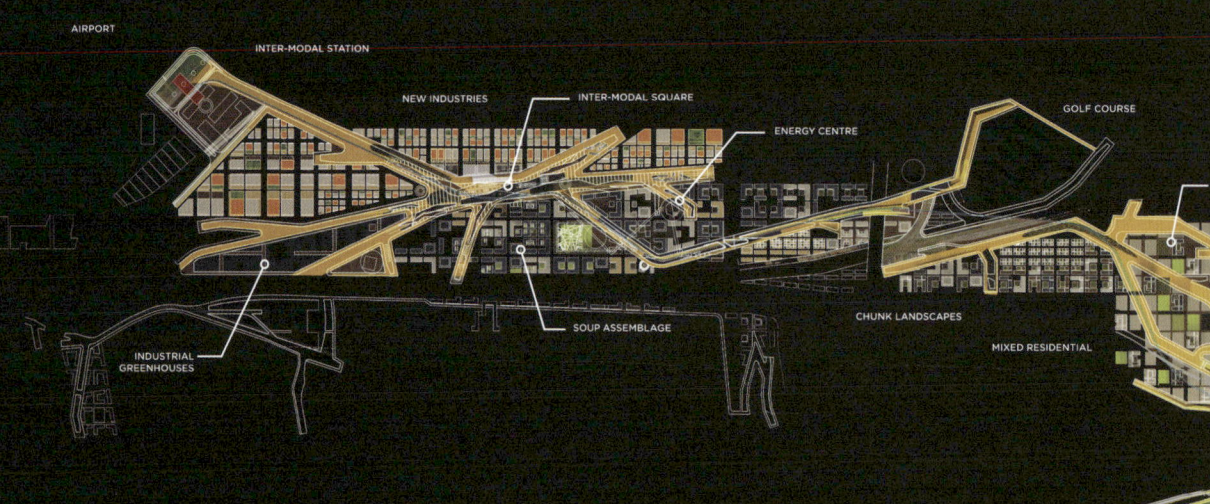

NEW CITY EDGE

CONNECTED WATER TREATMNET PLANT

LARGE SCALE SOUP ASSEMBLAGES

INTENSIFIED LACHINE WATERFRONT

LIFESTYLE CENTRE

NEW CHUNK NEIGHBOURHOODS

HOTEL AND INTER-MODAL STATION

NEW PATCH INDUSTRIAL

The 'chunks' are characterized through programs of open spaces and built form called 'soup assemblages' and 'filigree fogs'. These are made in turn by reconfigured industrial zones where sprawling areas are rationalized to be more compact and ecologically beneficial. They comprise shared infrastructure, large greenhouses, fields of solar collectors, dense residential mixes of high-rise and courtyard housing, and new entrepreneurial building types.

Also proposed is a transformation of the present suburban shopping mall (a large mass surrounded by asphalt) to a 'lifestyle mall', a city centre superimposing a larger framework of shopping galleries, courtyards, parking structures, and hotels. This straddles the expressway and connects the edges of the international corridor and adjacent neighborhoods. Similarly, the Lachine Canal, currently immobilized on its edges by a road and green fringe, is strengthened by making contact in less passive ways with new sites, frontage, and docking facilities, reviving it as a waterfront and an essential component of the new network's organization.

The 'strings' of Undercover Montreal are lines of intensity, which connect, ascend, descend, and bridge through the infrastructural network. Strings link neighborhoods to transit stations to cycling paths to lines of trees and linear parks, to shopping and work places. Strings, characterized as 'jets', 'lightning bursts', and 'mushrooms' are alignments of terraces, stairs, ramps, landscape forms, and transport nodes for buses, waterways, subways, and trains – creating, maintaining, and intensifying new relations.

The large-scale patchwork of 'chunks' provides a strategy of development for the international gateway corridor that has the potential to provide capital for the ongoing mirror development of Undercover Montreal. As large areas of formerly industrialized zones are rationalized into more compact and efficient zones, areas are freed up for residential and commercial development, which in turn frees funds and impetus for the infrastructural network. Again, this network, as it progresses, attracts further development.

Comments from the jury

This proposal stood out due to its in-depth study of the densification potential of areas adjacent to the highway. The submitted drawings suggest the use of a regular grid for the development of new urban blocks that are consistent with the shape of the existing neighborhoods. They also use a variety of colors to suggest the new development's multifunctionality and thus a mix of working, living, and recreational areas. The jury also appreciated the identification of an area in close proximity to the highway to insert large commercial structures ("life-style center").

The link that is made in this proposal between urban densification and infrastructure projects is seen as relevant because it points out a new funding strategy for infrastructure renewal, where the profit generated by new development could contribute to the funding of infrastructure projects. Thus, the infrastructure/neighborhood relationship would no longer be seen only in a functional perspective, but also an economical one.

The preparation of such an urban strategy could allow all local stakeholders involved in the planning of the gateway corridor to coordinate their actions, so that this territory becomes a priority area for development at the regional level.

Infra-sutures

dlandstudio / Trollback + Company, New York, United States
Susannah Drake, Main Partner, dlandstudio
Yong Kim, Partner, dlandstudio
Forbes Lipschitz, Designer, dlandstudio
Jakob Trollback, Creative Director, Trollback + Company
Rachelle Madden, Executive Producer, Trollback + Company
Erica Hirshfeld, Head of Production, Trollback + Company
Peter Alfano, Senior Designer and Technical Director, Trollback + Company

Figure 3.4: Planted "sutures" along with densification strategy for Autoroute 20 gateway corridor. (Excerpt from the proposal of dlandstudio, United States)

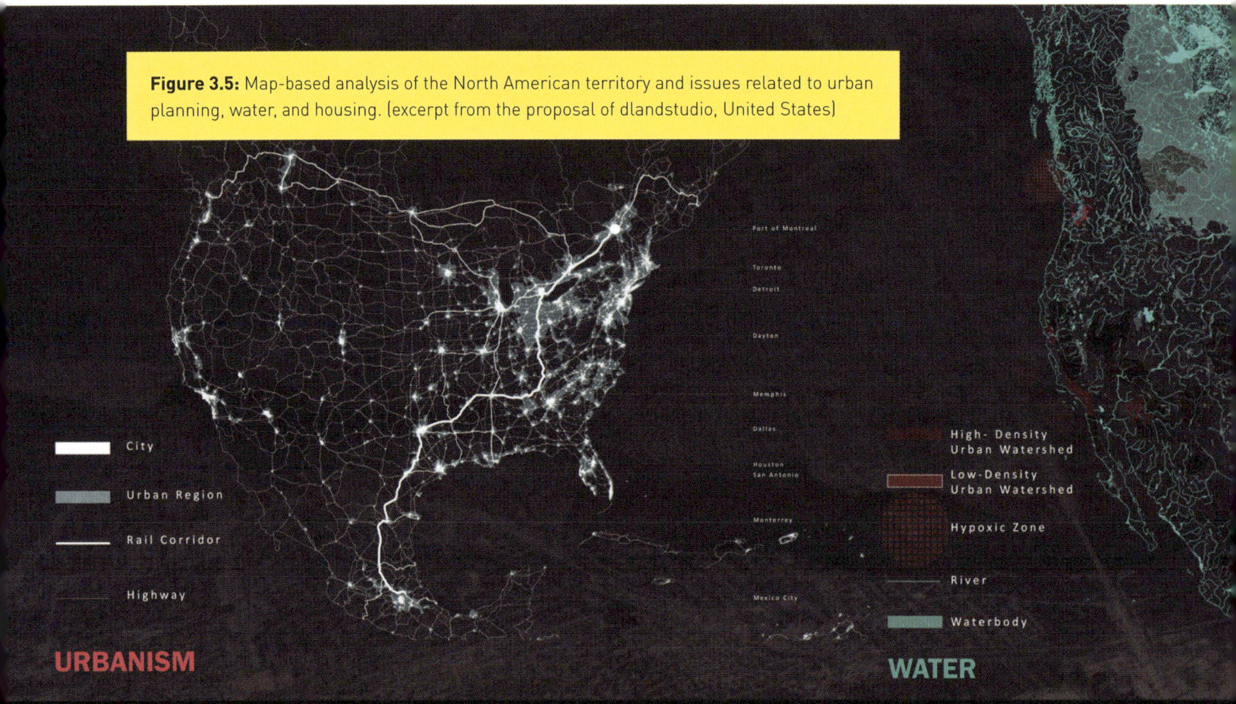

Figure 3.5: Map-based analysis of the North American territory and issues related to urban planning, water, and housing. (excerpt from the proposal of dlandstudio, United States)

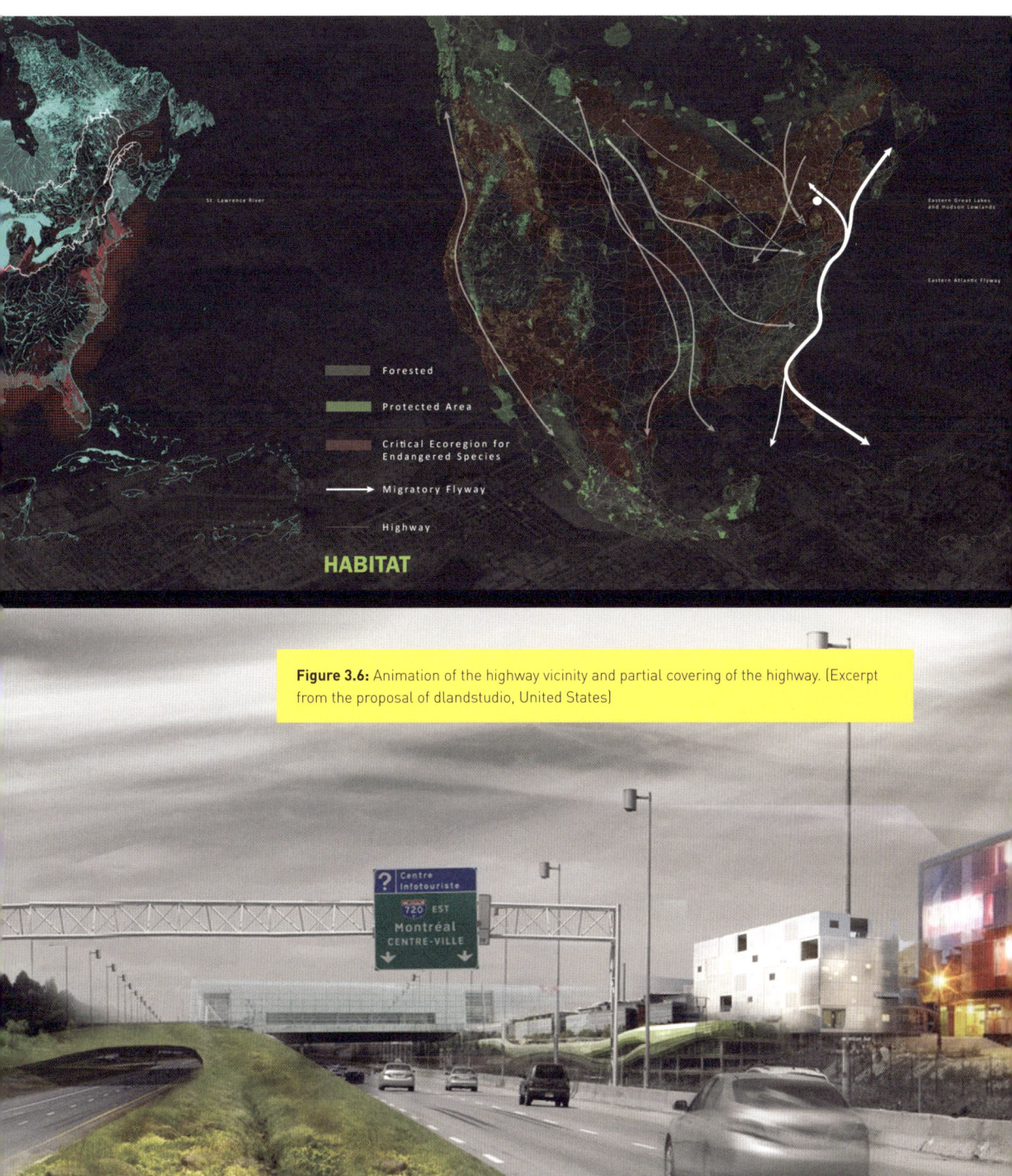

Figure 3.6: Animation of the highway vicinity and partial covering of the highway. (Excerpt from the proposal of dlandstudio, United States)

Text submitted by dlandstudio

Montreal is a city of design. This is not evident when arriving from Trudeau International Airport. What is apparent are the highways, canals, and freight lines that made the city a critical component of North American industrial trade. These specifically designed linear infrastructure systems are the dominant element in the landscape, yet they generally fulfill only a singular function. As a consequence, the urbanity, hydrology, ecology, creativity, and energy of the city are sublimated. Transit infrastructure built Montreal, but left it wounded. Streams are buried and flood plains paved. Open and urban spaces are dehumanized, disconnected, and dissonant with the image of the city and the surrounding environment.

The site is a microcosm of the North American infrastructure network, one of the largest open space systems in the continent.

Transit urbanizes. From Montreal to Mexico City, transit infrastructures define communities, towns, cities, states, and regions. This system enables the proliferation of low-density urban mega-regions across the continent.

Transit enables production. Agricultural and industrial economies rely on an expansive distribution network of rail and highway infrastructure. As the international port closest to North America's industrial heartland, the Port of Montreal represents a hinterland of some 100 million Canadian and American consumers.

Transit consumes energy. North America produces one-fourth of the global energy supply and consumes thirty percent of the world's commercial energy. As highways have enabled the proliferation of low-density urbanization across the continent, energy infrastructures have grown in both magnitude and complexity.

Transit pollutes. Through the proliferation of impervious pavement and the draining of wetlands, transportation infrastructures impede natural water purification processes. As a consequence, coastal dead zones expand in North American waters.

Transit fragments habitat. Current development patterns reduce habitat with a prevalence of pavement and buildings that interrupt natural systems. Though national agencies have attempted to manage the loss of naturalized areas through the national park system, the piecemeal pattern of protected open space cannot reconnect fragmented habitats.

As this infrastructure reaches the end of its design life, a new opportunity exists to create a system of **infra-sutures** that stitch together the urban landscape. **Infra-sutures** generate sustainable urbanism, support productive economies, enable alternative energies, restore natural hydrology, and reconnect habitats.

Operations that shift, span, submerge, and stilt transportation systems are the sutures. As pilings for new transport corridors are constructed they can go deep, becoming geothermal conduits that provide energy for new development and vertical agriculture. Stacked, high-density industry and specialized design manufacturing expands along the newly defined rail/habitat zones. Parks and linear green infrastructure systems stitch together neighborhoods, remediate pollutants, sequester carbon, and provide continuous habitat. New open space, coupled with the expansion of the Lachine Canal waterfront, provides a platform for high-density development. Green streets, blue roofs, and constructed wetlands absorb water and prevent runoff, while buildings in low-lying areas have freeboard to work with variable water levels. Infra-sutures change not only the productive capacity of transportation infrastructures, they create a new paradigm of experience for the city of the future.

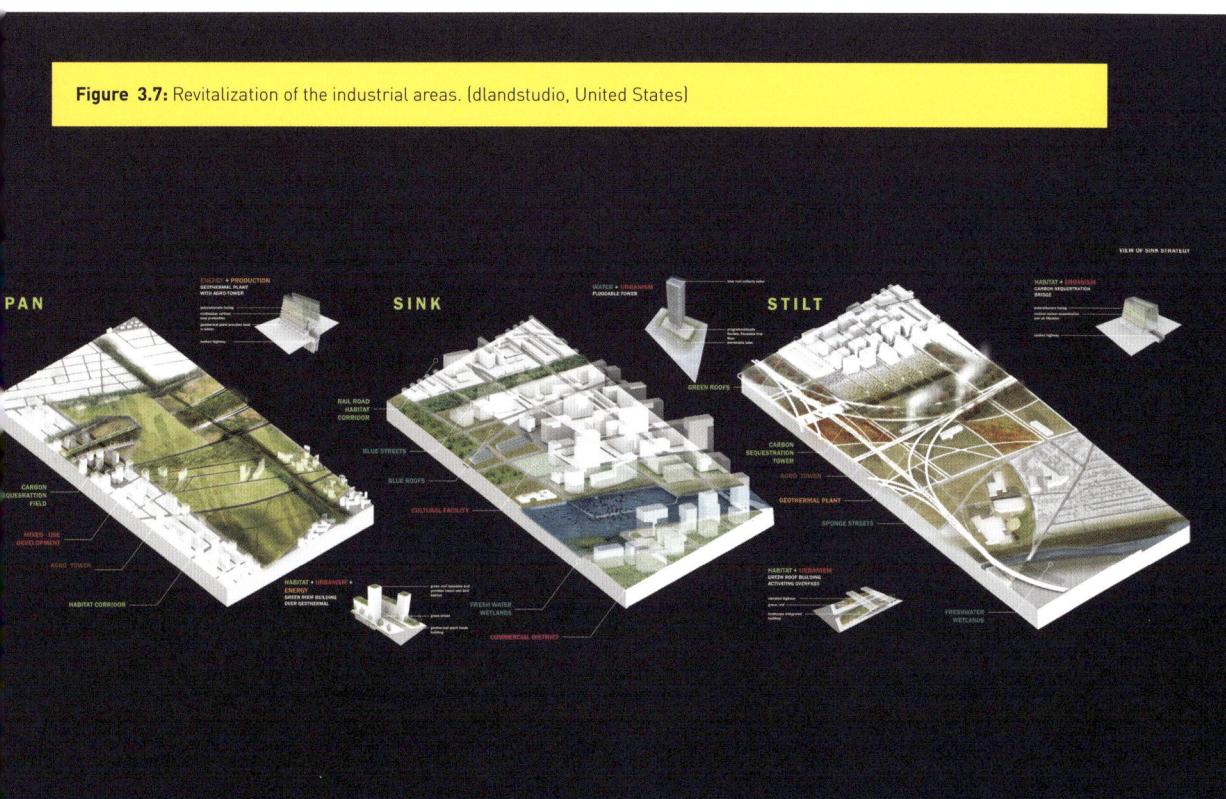

Figure 3.7: Revitalization of the industrial areas. (dlandstudio, United States)

Figure 3.8: Infra-sutures. (Excerpt from the proposal of dlandstudio, United States)

Comments from the jury

This proposal stands out through its elaboration of a coherent overall strategy, which considers the highway corridor as much as the adjacent territory. The contextualization of the highway within the North American highway network effectively illustrates the usefulness of the infrastructure while stressing the need to deal with the interconnection of adjacent neighborhoods. By remaining at the urban rather than architectural level, the proposal allows the development of a wide variety of strategies for the coherent articulation of several aspects of development, including public transportation, hydrology, natural habitats, urban development, and energy production. In this respect, the video demonstrates the overall challenges involved and communicates them properly.

One of the strong points of the proposal is the design of three types of north-south crossings of the highway: span, sink, and stilt. These three proposals for the highway crossing were not perceived as projects to be implemented, but rather as principles, whose combination can be adapted to the context, which gives a great deal of adaptability potential to the proposal.

The jury noticed that the proposal identified the potential for high-density development while freeing up green spaces. These new expressions of the adjacent environments are designed based on a work that is consistent with the existing urban grids, rather than by the construction of iconic landmarks.

Production <> Consumption

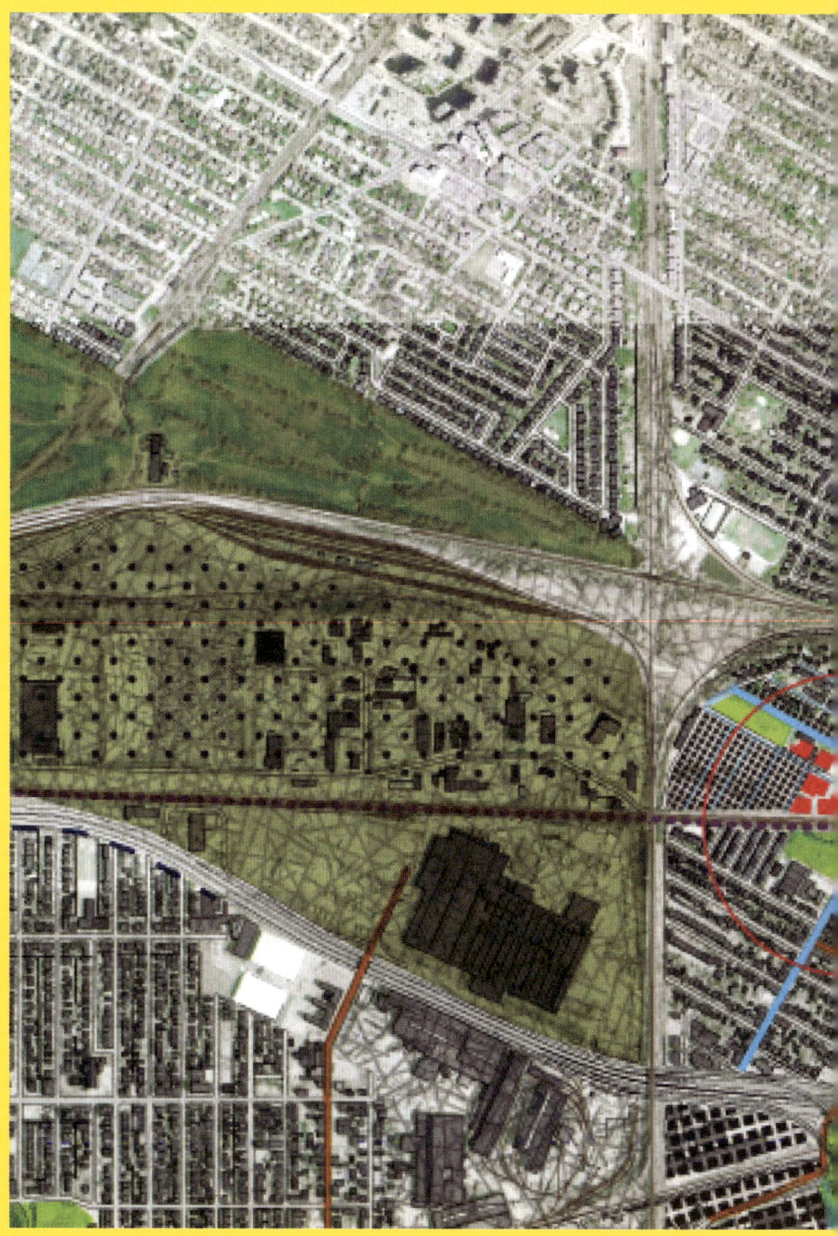

Gilles Hanicot, Montreal, Canada
Gilles Hanicot, Landscape Architect

YUL/MTL ILLUSTRATION

Figure 3.9: Urban planning strategy for the gateway corridor that redefines the industrial areas. (Excerpt from the proposal of Gilles Hanicot, Canada)

Figure 3.10: Public and emblematic art acting as a light source and allowing animation in the outskirts of the highway. (Excerpt from the proposal of Gilles Hanicot, Canada)

Figure 3.11: Using existing transportation lines and implementing new ones. (Excerpt from the proposal of Gilles Hanicot, Canada)

Text submitted by Gilles Hanicot

The *Production <> Consumption* project suggests to develop a **sustainable multifunctional urban program** that articulates around the implementation **of structuring and cautious infrastructure networks**, mainly thanks to **massive investments in public transportation**.

The project suggests, among other things, an important residential densification of the YUL/MTL corridor by converting industrial areas. The neighborhood would add attractive features that appeal to developers, with the aim of completely recasting the public transportation network and zoning by-laws, creating parks and green spaces throughout the highway course, establishing an express railway, etc.

***PRODUCTION <> CONSUMPTION* PUTS FORWARD THE IDEA THAT PUBLIC TRANSPORTATION CAN, AND SHOULD, SERVE AS A SKELETON FOR SUSTAINABLE URBAN DEVELOPMENT.**

Production <> Consumption expresses a complementarity, a dialog between the territory and the infrastructure. The territory and its intrinsic qualities are at the service of their infrastructure. Conversely, these public transportation infrastructures structure the territorial development from which they draw their energy. The infrastructures are at the service of their territory.

PRODUCTION <> CONSUMPTION – TERRITORY <> INFRASTRUCTURES

The project invests in marking territorial identity along the highway corridor. The establishment of infrastructure for renewable energy production becomes both a symbol of futurism and creativity for Montreal, and a source of energy for the transportation infrastructures.

The proposal thus identifies a new industrial activity that could be integrated into the highway corridor: renewable energy production. On top of reinventing the image of the corridor, the development of this industry would provide the energy necessary for the operation of local transportation infrastructures. To support this transformation, a collection of urban landmarks would be developed, such as lighting equipment.

TURCOT, ANGRIGNON, WEST-MONTREAL, ATWATER POLE AREAS

The interventions in these areas invest in a very strong urban densification by converting industrial spaces.

Angrignon Boulevard's course is detoured to alleviate the structures of the interchange and open up the Saint-Pierre district. The area point of articulation relies on a platform park above the highway. The park platform plays a major role, welcoming a new station that connects the neighborhood to downtown and the airport by implementing a pedestrian lane, bicycle lane, and a road toward the canal, the Turcot rail yard, and the southern areas.

Figure 3.12: Renewable energy production. (Excerpt from the proposal of Gilles Hanicot, Canada)

Connectivity between neighborhoods is ensured by implementing a coherent public transportation network; a large loop for an electric bus line on both sides of the canal.

The electric train, the backbone of the project, is an infrastructure at the service of its territory, a necessary and essential element for sustainable development of the urban skeleton; it provides shuttle bus service between downtown and the airport.

The Saint-Jacques Escarpment is reattached to neighboring districts through a platform park. The linear park of the canal is created along both the north and south side of the shoreline. The linear park creates a green axis that interconnects with the various neighborhoods. The appropriation of the surrounding area of the canal by new residential districts is strengthened by the creation of new parks adjacent to the linear park, which penetrate the residential areas and even sometimes connect to existing parks.

Norman

The Norman industrial area is a territory at the service of its infrastructure: the area is completely redefined for energy eco-production. The creation of a field with windmills and solar panels will be used to feed an electric train line, backbone of the corridor.

Saint-Pierre district

For this area, the main issue identified by the proposal is opening it up. To achieve this, the project suggests the following interventions: opening the district on the platform park of the new Turcot train station; creating a railway station on the YUL/MTL shuttle rail path; creating a neighborhood park, establishing a mixed and commercial area nearby; concentrating the residential spaces; redefining the north-south connectivity between the neighborhood and the linear park of the canal; and improving the connection between the neighborhood and the canal area by the extending Des Érables Street above the railway track.

Cabot

The area's development is provided by enhancing its industrial heritage and converting buildings for housing. Also suggested: opening on the canal and the linear park, connecting the West Turcot and downtown areas via a line of electric buses, concentrating and building housing units of significant height along the highway, enhancing the views of the downtown massif, creating a green zone buffer along the highway infrastructures, and opening up the district through local roads under the highway structure.

Comments from the jury

This proposal sets itself apart by identifying a new industrial activity that could be integrated into the Autoroute 20 corridor, which is renewable energy production. On top of reinventing the image of the pathway, the development of this activity would provide the energy necessary for the operation of local transportation infrastructures. To support this transformation, the proposal develops a collection of urban landmarks, such as lighting equipment.

    The jury also stresses its interest in the suggested development phases for planning the corridor. These development phases implement a joint program that is structured into three sections, two of which are developed in the proposal: the first insists on the presence of an industrial park dedicated to renewable energies in the Norman Street area, and the second develops a high-density urban neigborhood in the area of the former Turcot rail yard and Angrignon Boulevard. The feasibility of the suggested ideas for the implementation of this territorial redevelopment program explains the jury's interest in this proposal.

YUL/MTL international ideas competition mentions

Andrew Forster – Push Montreal: Montreal, Canada
Catalyse urbaine, architecture and landscape: Montreal, Canada
Clement Boitel: Paris, France
Dennis A. Winters, Tales of the Earth: Toronto, Canada
Efoe Arnaud: Clermont-Ferrand, France
Superlandscape: Padova-Palermo, Italy
Gerwin De Vries + Alexander Herrebout: Utrecht, The Netherlands
Ghazal Jafari and Ali Fard: Toronto, Canada
Thibodeau, architecture+design: Montreal, Canada
Yvette Vasourkova, MOBA Studio: Prague, Czech Republic
zerOgroup + Fabrica de Paisaje: Sao Paulo, Brazil

Figure 3.13: Audio perception of the characters of the pathway. (Excerpt from the proposal made by Andrew Forster, Push Montreal, Canada)

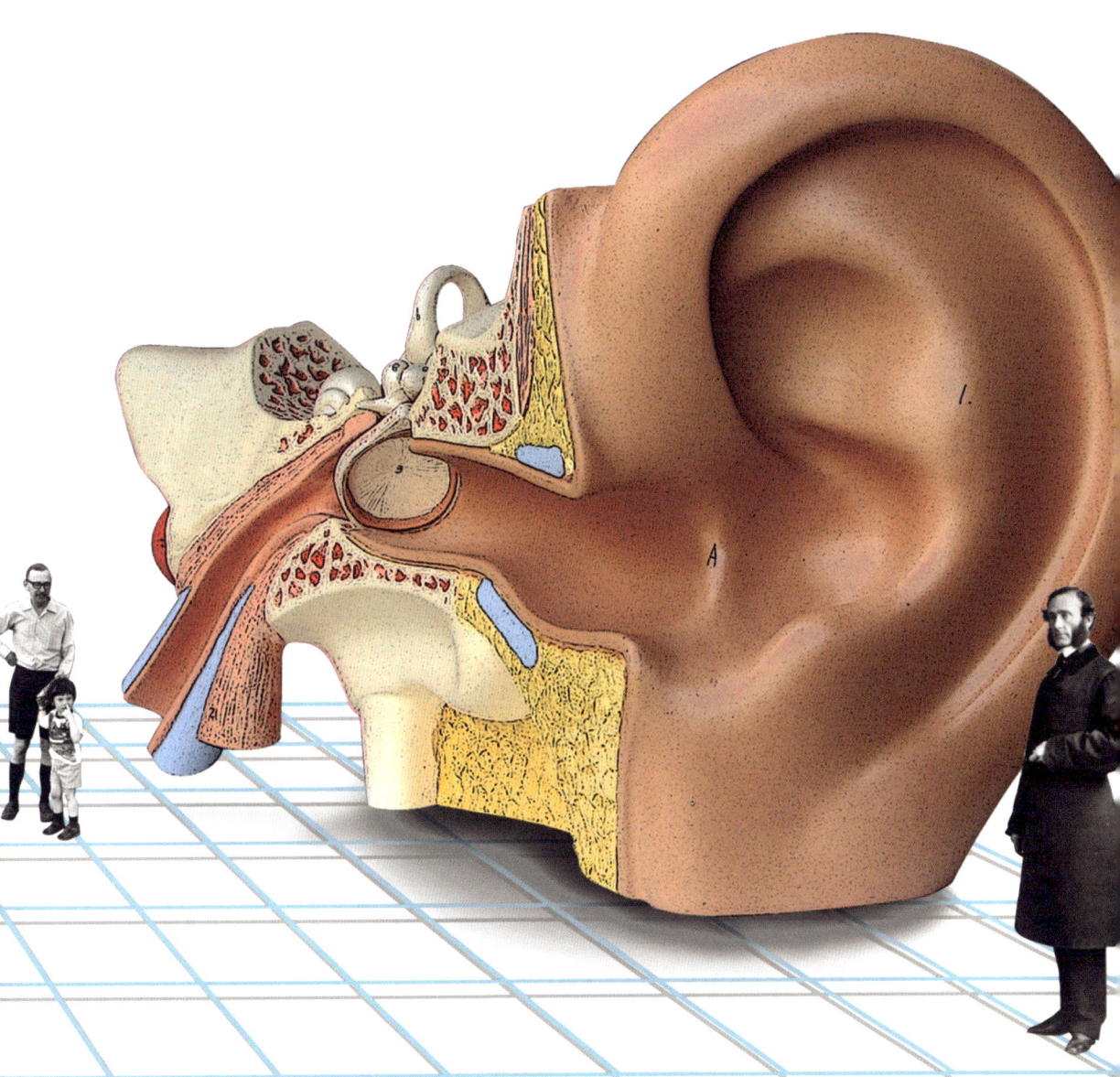

Figure 3.14: Inhabited bridge, whose wall allows signage. (Excerpt from the proposal of Ghazal Jafari + Ali Fard, Canada)

Figure 3.15: Vegetation along the highway right-of-way. (Excerpt from the proposal made by Superlandscape, Italy)

Figure 3.16: Improving the airport/downtown connection. (Excerpt from the proposal made by Efoe Arnaud, France)

Figure 3.17: Modifying the topography of the gateway corridor and enhancing Mount-Royal. (Excerpt from the proposal made by Gerwin de Vries + Alexander Herrebout, The Netherlands)

Figure 3.18: Using an industrial artifact of the Montreal identity, the container, to animate the gateway corridor. (Excerpt from the proposal of Clement Boitel, France)

Figure 3.19: Implementing emblematic buildings in the highway right-of-way. (Excerpt from the proposal made by ZerOgroup + Fabrica de paisaje, Brazil)

Figure 3.20: Creating an ecological and recreational corridor. (Excerpt from the proposal made by Moba Studio, Czech Republic)

Workshop_Atelier/Terrain UNESCO Montreal 2011

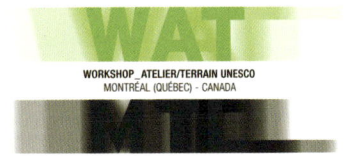

The Workshop_Atelier/Terrain (WAT_UNESCO) of the UNESCO Chair in landscape and environmental design of the University of Montreal (CUPEUM) is an international activity supported by UNESCO since 2003, held annually in different cities around the world. This annual meeting provides an opportunity to reflect on the future of cities, and is supported in Montreal by five UNESCO programs:

MOST (Management of social transformations)
Creative Cities Network
EDD (Education for a sustainable development)
MAB (Man and the Biosphere)
IHP (International Hydrological Program)

In Montreal in 2011, the WAT_UNESCO presented its eighth edition. The previous editions took place in the cities of Reggio de Calabria (Italy 2003), Marrakesh (Morocco 2004), Saïda (Lebanon 2005), Mahdia (Tunisia 2006), Ganghwa (Republic of Korea 2007), Jinze-Shanghai (China 2008), and Kobe (Japan 2009)[1].

As an ideation exercise, each edition invites CUPEUM's international network to develop strategic urban design visions. This activity also allows exchange and international networking between upcoming urban design professionals (architecture, landscape architecture and urban planning), experts in urban planning, and local and governmental officials of the host cities. WAT_UNESCO stimulates the implementation of imaginative and innovative projects for the livable development of cities of the South and the North.

The general purposes of WAT_UNESCO intend to carry international strategic reflections on the planning and development of landscapes and urban living environments through a perspective of sustainable development, which is to:

---

[1]. To view the themes and projects of WAT_UNESCO, visit: (http://unesco-paysage.umontreal.ca/fr/recherches-et-projets/workshop-atelier-terrain-wat-unesco)

- Characterize the major territorial issues, attractive features, and potentials of studied urban landscapes;
- Provide structuring and viable planning solutions to governments and territorial stakeholders by carrying out tailored landscape projects addressing conservation, enhancement, and development challenges;
- Generate a multidisciplinary reflection and transdisciplinary knowledge between professionals and scholars (architects, landscape architects, urban planners, engineers, sociologists, historians, geographers, etc.) and include the activity's results in CUPEUM's *International observatory of urban landscapes: cities and metropolises*.

WAT_UNESCO Montreal[2] brought together forty-eight students and twelve professors from twelve different academic institutions, making it an unrivaled apprenticeship model through its multicultural and plural approach toward involved knowledge and its sustained dialogs between all stakeholders (elected officials, experts, citizens, scholars, etc.). Integrated into the *YUL/MTL: Moving Landscapes* project, this edition was jointly supported by the City of Montreal and the ministère des Transports du Québec (Canada).

The institutions participating in WAT_UNESCO Montreal 2011 were:

University of Montreal (Faculty of Environmental Design) - Canada
Tongji University (College of Architecture and Urban Design) - China
University of design of Kobe (School of Architecture) - Japan
University of Kobe (Department of Architecture) - Japan
Meiji University (Architecture department) - Japan
Tunis National School of Architecture and Urban Planning - Tunisia
American University of Beirut (School of Landscape Architecture) - Lebanon
Lebanese University (Institute of Fine Arts) - Lebanon
National School of Architecture - Morocco
University of Rome La Sapienza (Department of Architecture) - Italy
University of Sassari (Department of Architecture of Alghero) - Italy
Damascus University (Department of Architecture) - Syria

By producing alternative visions of urban planning on a local scale (micro-design), inspired by the winning visions of the *YUL/MTL: Moving Landscapes* international ideas competition, participants in WAT_UNESCO helped to enrich the dialog process between stakeholders of the local territorial development, as well as to illustrate the potential of a shared intervention strategy to facilitate appropriation.

---

2. For more information about this workshop, visit: (http://www.unesco-paysage.umontreal.ca/fr/recherches-et-projets/workshop-atelier-terrain-montreal-2011)

Intervention site 1: Dorval

The beginning or ending point of Montreal's international gateway corridor, the Dorval site is composed of several transport networks, whose scope covers several territorial scales. At the interurban and regional scales, mainly articulated by the air terminal, the Via Rail passenger station, Autoroutes 20 and 520, as well as the AMT suburban trains station, the pathway is unidirectional and aims at providing a direct connection between downtown and this sensitive transportation node. Here, the urban planning challenges concern the symbolic nature of the perceived landscapes to display the identity of Montreal.

At the local scale, with the presence of the bus terminal and the local arterial network, the paths are multi-directional and aim to access different neighborhoods and points of interest of Dorval. The urban planning challenges are multiple and concern both the interconnection between different transportation equipment and the harmonization of the territorial development with opportunities offered by the large transportation supply, particularly through the creation of a Transit-Oriented Development (T. O. D.).

Intervention site 2: Lachine

An expression of a corridor rationale, the transportation network's right-of-way (highway and railway track) on this site causes an important territorial division: first functional, with the separation of Lachine's residential and industrial areas, and then physical, by the creation of a barrier through which crossings are almost non-existent.

For their part, the north and south sides of the corridor animate the entrance paths with their own expressions, with noise-reducing walls on the south side and natural wastelands on the north side. In their own way, they each filter the perception from the adjacent environments and vice versa, conditioning the opportunities of the urban context to adapt to their presence.

For the highway component, the meeting point of Autoroute 13 and 32nd Avenue with Autoroute 20 constitutes an important articulation of the entrance pathway. For its part, the railway network also has an anchor point through the Lachine railway station and the pedestrian tunnel, allowing access at 48th Avenue.

Intervention site 3: Saint-Pierre

Structured around the Saint-Pierre interchange, this site represents a sensitive intersection for transportation in Montreal. At the intersection of two important highway axis, A-20 and Road 138, the Saint-Pierre interchange is also the meeting point of four local roads: Saint-Jacques Street, West Notre-Dame Street, Saint-Joseph Boulevard, and Saint-Patrick Street. It also represents an intersection for other transportation infrastructures, such as the CN railway, the CP railway, and the Lachine Canal. In addition to causing significant problems for using the interchange, this juxtaposition of transportation lines restricts the accessibility between neighborhoods and industrial areas. The reconstruction of the Saint-Pierre interchange brings with it the possibility of rethinking the management of highway traffic, the relationship between the neighborhoods and the infrastructure, and the image of the infrastructure itself. Could the upcoming road project be the main vehicle for designing a new urban project? That is the challenge of the Saint-Pierre site.

Intervention site 4: Former Turcot rail yard

An industrial wasteland on its way to being completely redeveloped, this vast territory offers the greatest opportunity for reinventing the landscapes of the gateway corridor that connects the Montreal-Trudeau Airport to downtown. With the presence of industrial artifacts, as well as the Saint-Jacques Escarpment and the Lachine Canal, this area has a high potential for enhancing the local natural and industrial heritage, as well as redefining the relationship between transportation infrastructure and its adjacent environment.

Intervention site 5: Cabot and Côte-Saint-Paul

Built on elevated structures, this section of Autoroute 15 offers a remarkable, continuous view of the Montreal skyline and Mount Royal. The rehabilitation of this infrastructure maximizes the enhancement of this view and highlights its main features with signal elements, like when going over the Lachine Canal, or with a high quality civil engineering structure.

At ground level, we find urban areas heavily affected by the nuisance caused by the highway. Yet here, reflection on the renewal of highway equipment allows implementing the catalysts of various revitalization efforts for the urban and industrial neighborhoods located on both sides of the highway.

Intervention site 6: City center access corridor

This site juxtaposes highway and rail infrastructures in the same corridor. Their implementation on different levels of the Saint-Jacques Escarpment allows a continuous view of the roofs and church towers of the lower city, as well as an axial perspective on the towers of downtown.

Although the pathway has exceptional views, some of the areas are undervalued and show low-quality public spaces. In addition to the insertion of highway entrances/exits, the undervaluation of adjacent areas raises issues related to the nuisances caused by the highway and the layout of the highway's edge.

AREAS
1. DORVAL
2. LACHINE
3. SAINT-PIERRE
4. FORMER TURCOT RAIL YARD
5. CABOT AND CÔTE-SAINT-PAUL
6. CITY CENTER ACCESS CORRIDOR

LEGEND
- URBAN FABRIC
- RAILWAY PATH
- DRIVING TOUR
- STATION / AIRPORT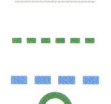
- INTERCHANGES / POINT OF VIEW
- PRIORITY PLANNING AREA
- SECONDARY PLANNING AREA

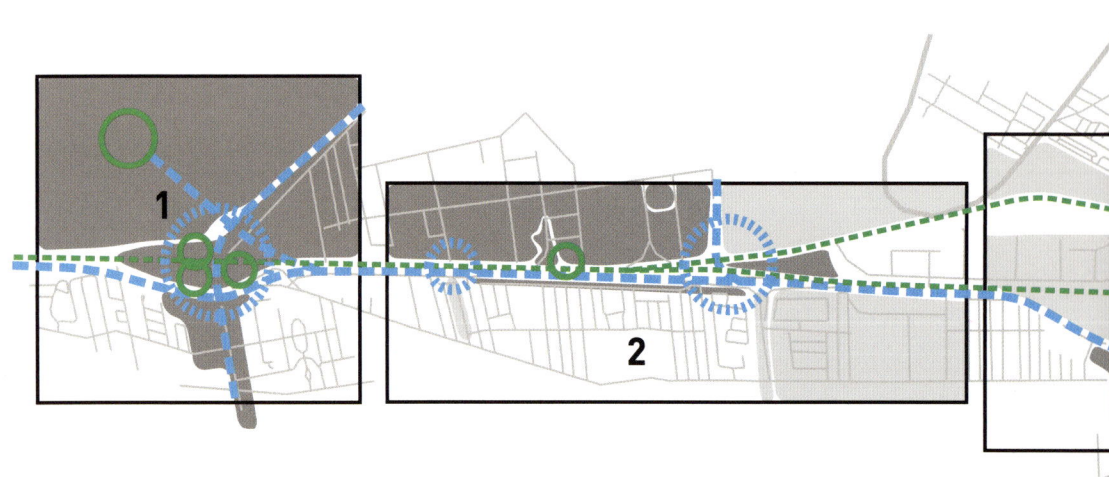

YUL/MTL ILLUSTRATION

Figure 3.21: WAT_UNESCO Montreal intervention sectors and opportunity sites.

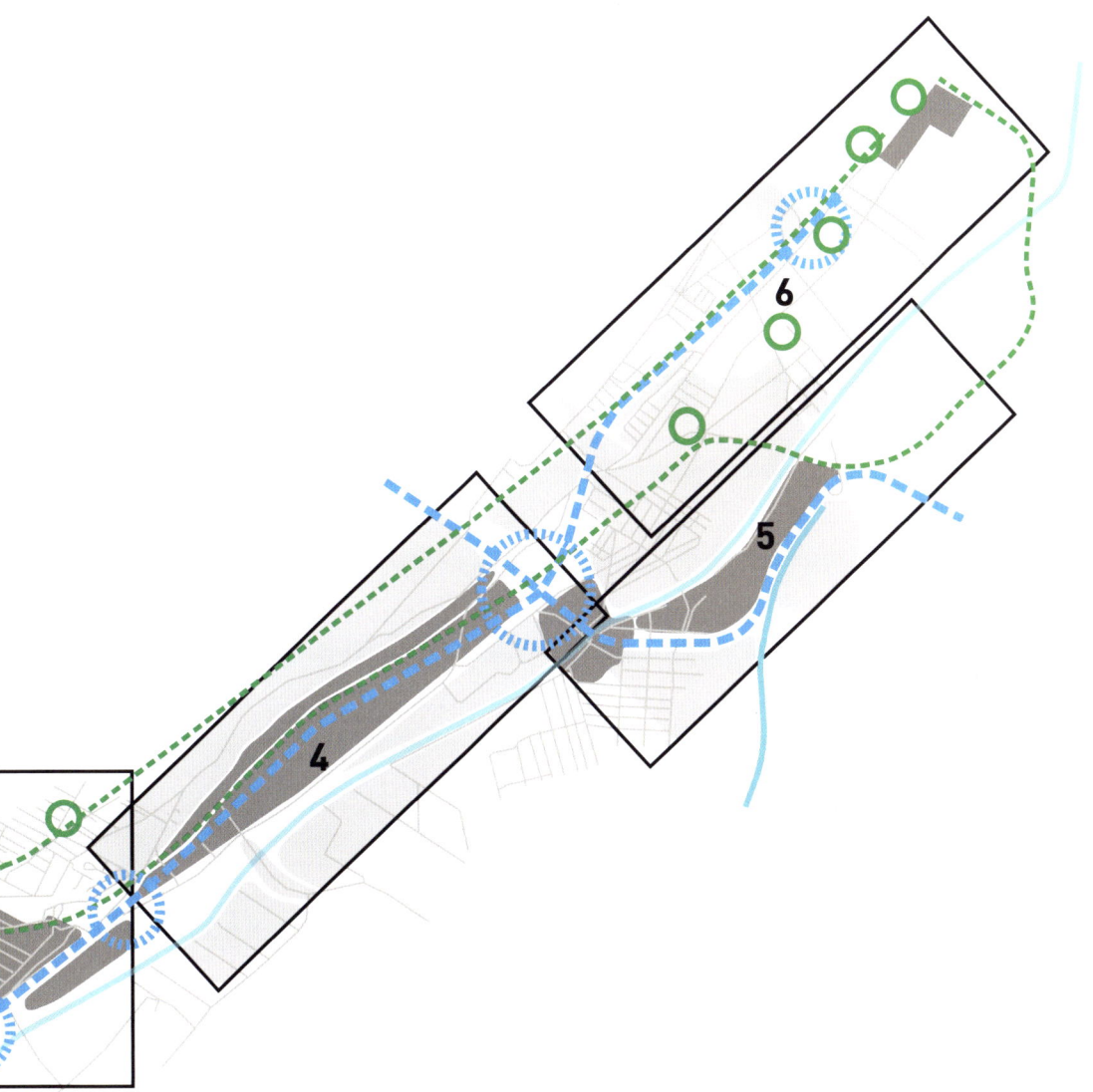

Areas of intervention

1. Dorval
2. Lachine
3. Saint-Pierre
4. Former Turcot rail yard
5. Cabot and Côte-Saint-Paul
6. City center access corridor

Related issues

Identity

Identity character of Montreal's international city entrance

Expression Noise barriers

Aesthetic quality of civil engineering structures

Quality of public spaces and infrastructures

Pathway

Experiences of entry/exit of the Montreal-Trudeau Airport

Experience of infrastructure by local residents

Interface between the highway and adjacent areas

Sustainability

Reduction of the Highway barrier effect for adjacent areas

Enhancement of residual spaces

Enhancement of natural and recreational areas

Quality of life of adjacent areas

Requalification, optimization of industrial land

Vocation of spaces around the terminal

YUL/MTL   ILLUSTRATION

Description of student projects

Site 1 / Dorval

THINK GLOBALLY, LIVE LOCALLY

The main idea of the *Think globally, Live locally* proposal is to fill and animate empty and residual spaces resulting from the highway infrastructure, particularly in locations where access ramps are numerous. With the Dorval interchange being the subject of the study, issues of welcoming, transit, threshold, and identity were mainly addressed.

A proven urban planning solution is that of an architectural ribbon with a variable topography that combines the functions and intervention scales. In this sense, the ribbon participates in creating an expressive visual landmark for the city entrance and integrating the adjacent residential neighborhoods into a network of friendly spaces. At the global scale, the relevance of this megastructure depends on the improvement of access to public transportation and the creation of natural corridors (vegetation, waterways) that pass through the highway. At the local level, the insertion of landscaped ditches and the plantation of trees, included in this new structure and the residential urban fabric, can improve the quality of life of the neighborhoods.

TEAM A: *Andrea Becca from the University of architecture of Alghero (Italy), David Fiset from the University of Montreal (Canada), Natsuko Kobayashi from the University of Kobe (Japan), and Zeineb M'Rabet from the National School of Architecture and Urban Planning (Tunisia), under the faculty guidance of Natalia Atfeh, University of Damascus, Syria, and Hiroyuki Sasaki University, Japan.*

Figure 3.22: New viewpoint from the airport.

Figure 3.23: Megastructure of the Dorval Interchange.

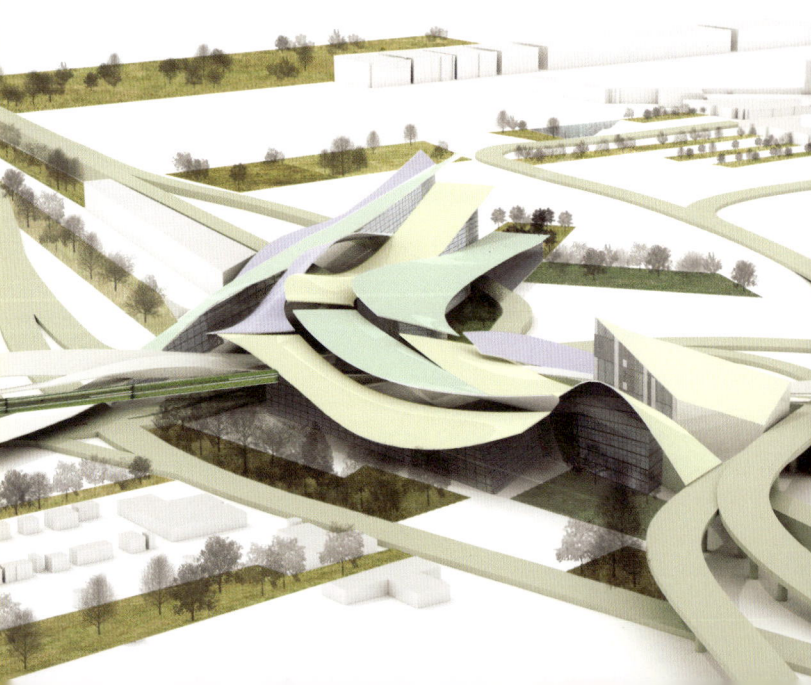

ID-ENTITY

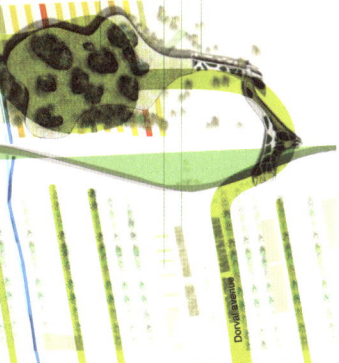

The complexity and the number of access ramps of Autoroute 20 took apart living environments, particularly in the case of the Dorval interchange. Based on this observation about the impacts of highway infrastructures, the *Id-entity* proposal deals with territorial challenges through three key themes: connection, vegetation, and transportation. The retained planning solution favors the development of ecological neighborhoods that improve the collective transportation lines. The transportation axes promote connections at different levels.

The project also suggests the realization of a long, elevated platform, which originates in the residential neighborhoods, goes over the highway, and extends into the vastness of the airport parking. This bridge also plays the role of a station, on which it is suggested to plant a park-like biomass. At the local level, this intervention expands and completes the neighborhood promenade network, as well as the connections between the existing parks. At the scale of the city of Montreal, it acts as an identity landscape and transmits the image of an ecological city geared toward sustainable development.

TEAM B: *Laurence Aubin-Steben from the University of Montreal (Canada), Mohamed El Tayeb from the National School of Architecture of Rabat (Morocco), and Shen Jiping from Tongji University (China), under the faculty guidance of Natalia Atfeh, University of Damascus, Syria, and Hiroyuki Sasaki, Meiji University, Japan.*

Figure 3.24: Elevated platform that connects the northern and southern parts of the Dorval district.

Figure 3.25: Strategy for regional development of the Dorval Interchange and its surroundings.

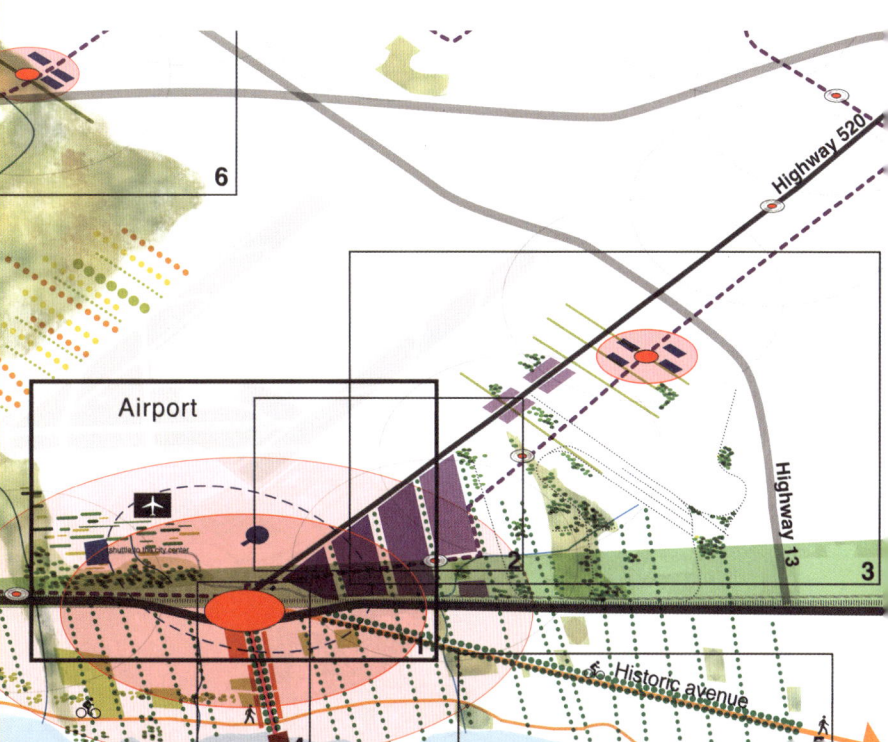

Site 2 / Lachine

TIME-LINE

The Lachine area is composed of a residential neighborhood, dilapidated industrial buildings, and an important rail yard. The *Time-Line* proposal suggests answering this particular situation with an ambitious urban design project.

The main urban planning solution is to create a link beyond the highway border, between the northern industrial area and the southern heart of the old district of Lachine. This departure intention comes in two actions: the creation of a North-South Lachine Boulevard and the gradual transformation of the rail yard (northern part) to agricultural, then urban areas. Ultimately, these interventions aim to improve the local quality of life by designing two poles: one in South Lachine, with a historical character facing the river, and one in North Lachine, focusing on diversity and innovation. Between these two poles, an inhabited structure crosses the highway and is viewed as the first residential development vector. The development of a visual interface to the highway is also designed separately, in light of the various adjacent environments' own specific features. It takes the form of a plant-based noise barrier in the northern section and a wall constructed from on-site materials (e.g.: brick).

TEAM A: *Bruna Bajramovic from the University of Design in Kobe (Japan), Paulo Casu from the University of Architecture of Alghero (Italy), Stephany Khoriaty from the American University of Beirut (Lebanon), and Amayel N'Diaye from the University of Montreal (Canada),* under the faculty guidance of Naoko Kuriyama, University of Kobe, Japan, and Leon Telvizian, Lebanese University, Lebanon.

Figure 3.26: Phase 2: Gradual transformation of the rail yard.

Figure 3.27: Noise-reducing brick wall that takes the same materiality as the adjacent neighborhoods.

Figure 3.28: Phase 3: Development of an urban environment that anchors in the new boulevard.

SNAKES AND LADDERS

Snakes and ladders emphasizes first the improvement of the gateway corridor's environmental qualities, with the aim of increasing the communities' quality of life. The proposal's purpose is to connect ecoterritories, reconstruct historic waterways, and develop quick public transportation that promotes the emergence of transit-oriented development.

Locally, this strategy promotes the development of green corridors that are perpendicular to the highway – avenues and inhabited bridges with diverse and lively programming that facilitate the highway's crossing. In the Norman yard, a green urban park enhances the industrial artifacts. In the South Lachine neighborhood, a new electrical train allows renovation of the neighborhood and creates a medium-density residential area.

Through redefining the Lachine area, the proposal also shows a constant desire to create interesting visual experiences both from the highway and toward adjacent territories.

TEAM B: *Linda Fortin from the University of Montreal (Canada), Salma Jarbi from the University of Damascus (Syria), Wiem Mami from the National School of Architecture and Urbanism (Tunisia), and Masamitsu Tanikawa from Meiji University (Japan), under the faculty supervision of Naoko Kuriyama, Kobe University, Japan, and Leon Telvizian, Lebanese University, Lebanon.*

Figure 3.29: Contemporary urban park that displays industrial artifacts and other relics from the past.

Figure 3.30: Multifunctional public space with iconic buildings.

Figure 3.31: Comprehensive plan of the Lachine area and neighborhoods oriented towards public transportation.

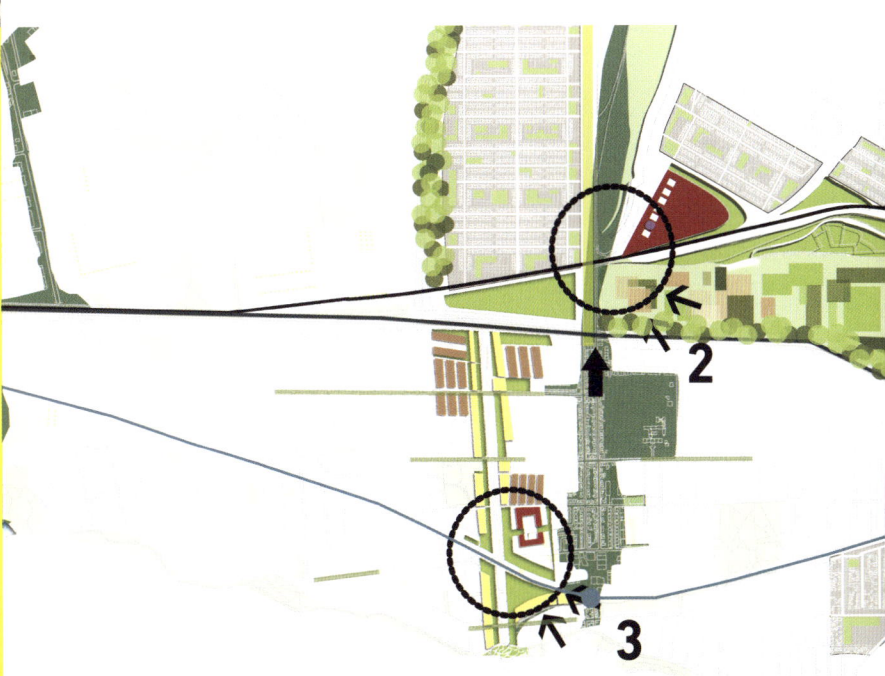

Site 3 / Saint-Pierre

HOT SPOTS

The intent of the *Hot spots* proposal is to highlight three features of Montreal's identity: its institutions, its culture, and its nature. The stance of the project relies on reading the opportunities offered by the territory through its existing infrastructures, particularly those of the rail network.

The findings of the global and local analysis result in a design strategy that suggests, through a 'zipper' analogy, to reunite neighborhoods while enhancing the linearity of the city entrance, to absorb the impact of the highway infrastructure, to create a plant biomass, and to develop attractive areas (Hot Spots). The mix of uses and activities relies on the economic, social, and cultural redeployment of the area.

For example, a lack of welcoming infrastructures at the intersection of two passenger train rail tracks turns into an opportunity to develop an intermodal station with an expressive architecture and innovative design. At the local level, the development of numerous artist studios facilitates the creative development of the new district.

TEAM A: *Ahmad Almahairy from the University of Damascus (Syria), Claudio Mura from the University of Architecture of Alghero (Italy), Christine Robitaille from the University of Montreal (Canada), and Bian Xiaozhe from Tongji University (China), under the faculty supervision of Alessandra Capuano, University of Rome La Sapienza, Italy, and Takeo Kawakita, Kobe Design University, Japan.*

Figure 3.32: Public space at the artists' pavilion entrance.

Figure 3.33: Natural and cultural planning strategy of the Saint-Pierre area.

Figure 3.34: Forecourt of the intermodal station under the Saint-Pierre Interchange.

GREEN SEA

Montreal is crossed by various transportation infrastructures (Lachine Canal, railways, motorways), along which factories were implemented that are now, for the most part, in a dilapidated condition, particularly in Saint-Pierre. Based on this analysis, the Green sea proposal puts forward two main interventions: that of using the brownfield potential to generate a major transformation across the city and that of exploiting the different characteristics of neighborhoods that have, in a context of isolation, forged different identities over time.

Specifically, the main idea of the project is to exploit the railway belt, which is losing economic vitality around Mount Royal, to imagine a new pattern of public transportation. The development of this new transportation network and intermodal stations allows a better connection between the cultural and event centers, creating unique territorial opportunities and enhancing Montreal's various identities. An ecological and social vocation within this broad corridor leaves space for vegetation, biodiversity, agriculture, and various outdoor activities, and improves the residents' quality of life.

TEAM B: *Reve Aoun from the Lebanese University (Lebanon), Caroline Cagelais from the University of Montreal (Canada), Valerie Gravel from the University of Montreal (Canada), Kohei Kobayashi from Kobe University (Japan)* under the faculty supervision of Alessandra Capuano, University of Rome La Sapienza, Italy and Takeo Kawakita, Kobe Design University, Japan.

Figure 3.35: Implementation of water taxis on the Lachine Canal and a boat station close to the new residential development.

Figure 3.36: Industrial brownfield potential, in which floats the different island-neighborhoods that have multiple identities.

Figure 3.37: Operation of the railway belt to connect the Saint-Pierre area via a new public transportation network.

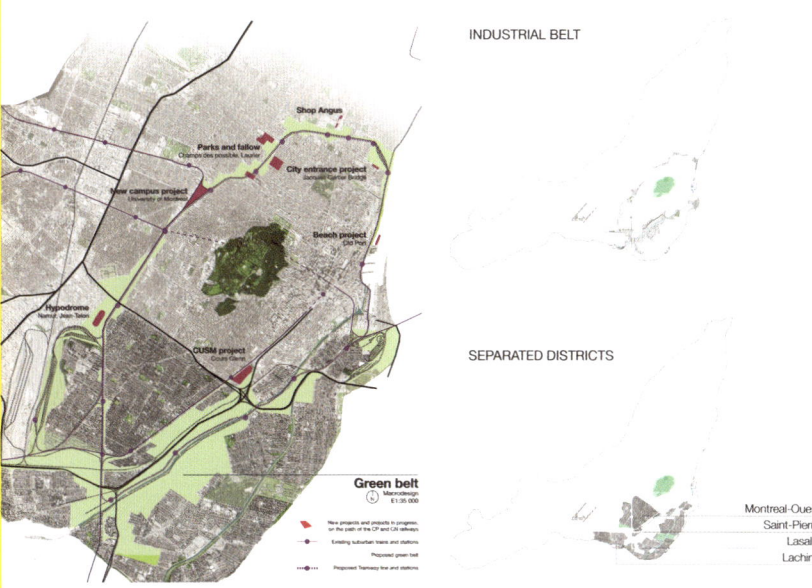

Site 4 / Turcot rail yard

MOVING CONTAINERS

The *Moving containers* project utilizes the heterogeneity of the Turcot area to establish the foundation of its proposal, such as the industrial past of the former rail yard, the ecoterritorial wealth of the Saint-Jacques Escarpment, and the presence of two transportation vectors from two eras, the Lachine Canal and the highway infrastructure. The main intention of the project is to connect these different environments through the use of a representative element of this industrial era: the container. Containers serve as a link between neighborhoods and benefit from their locations by creating new views or having an appropriate programmatic function.

In this sense, the stance of the project is to exploit the full potential of the Turcot yard beyond the escarpment, and even beyond the Lachine Canal, by establishing a new urban fabric. The dense, flexible, and evolving urbanization strategy, permitted by repeating the same module (e.g. container), also relies on interstitial green spaces and a phytoremediation technique

TEAM A: *Farah Aremmaz from the National School of Architecture of Rabat (Morocco), Julien Bourque from the University of Montreal (Canada), Rouba Dagher from the American University of Beirut (Lebanon), and Eri Ohara from Meiji University (Japan)*, under the faculty supervision of Salma Hamza, National School of Architecture and Urbanism, Tunisia, and Stefan Tischer, Faculty of Architecture of Alghero, Italy.

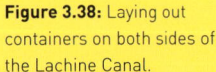

Figure 3.38: Laying out containers on both sides of the Lachine Canal.

Figure 3.39: Interstitial natural space of the new urban grid of containers.

Figure 3.40: Bridge over the Lachine Canal.

YUL/MTL

INTERLACE VALLEY

Many changes stemming from human action shaped the former Turcot rail yard site as we know it today. The *Interlace Valley* proposal recognizes the history of the site and, through the interweaving theme, suggests superimposing a new environment onto the existing natural matrix. This interweaving metaphor is present in both the planning of the site functions, as well as the built and natural elements articulated by the urban planning of adjacent communities. This strategy aims to create strong, lasting anchors for the Turcot area in the context of long-term planning.

A "green" and "blue" links concept implements a public space that integrates ecological principles, like a rainwater collection system, at a larger scale. This matrix also has a concern of use flexibility and is in the spirit of scripting heritage elements. Different corridors and poles of residential development are reflected in terms of social issues (art, work/community, health). Across the city, new transit lines connect emerging neighborhoods to the greater city fabric.

TEAM B: *Catherine Blain from the University of Montreal (Canada), Taiki Fujimaki from Kobe Design University (Japan), Valerie Poggiani from the University of Rome La Sapienza (Italy), and Andrea Spector from the University of Montreal (Canada)*, under the faculty supervision of Salma Hamza, National School of Architecture and Urbanism, Tunisia and Stefan Tischer, Faculty of Architecture of Alghero, Italy.

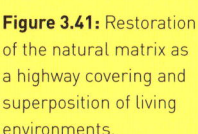

Figure 3.41: Restoration of the natural matrix as a highway covering and superposition of living environments.

Figure 3.42: Management strategy of surface waters.

Figure 3.43: Turcot neighborhood and links to the Saint-Jacques cliff.

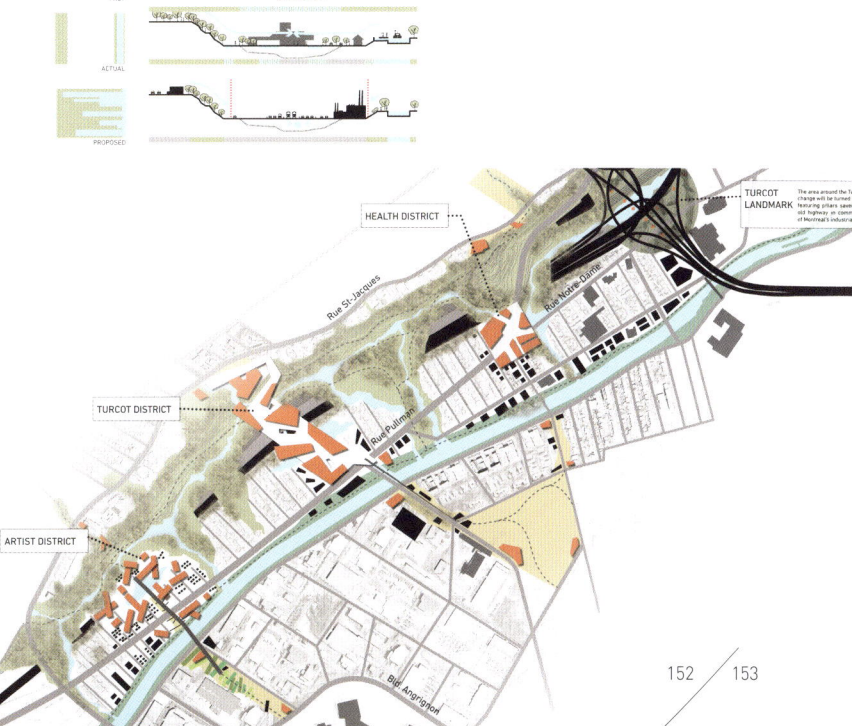

Site 5 / Cabot / Côte-Saint-Paul

IDENTITY PARK

The *Identity park* project roots from the site's industrial identity and promotes the idea of redefining the highway infrastructures with a particular function and meaning toward the territory. In this sense, urban infrastructures and adjacent urban areas play leading roles in the spatial deployment of the Cabot/Côte-Saint-Paul site. The proposal especially counts on an assemblage of urban grids, a public spaces network, and the creation of new morphologies to revitalize the area. Increasing the supply and diversity of transportation means across the city, as well as improving Montrealers' quality of life through better access to water bodies and natural spaces, are also issues at the heart of this proposal.

Identity is also a primary concern of this proposal. In this sense, industrial artifacts (e.g. water and tank containers) and interfaces are used to express the unique identity of areas and participate in the experience of the city entrance.

It is clear that the proposed design carries within it a very strong idea of opening up and the desire to assert a distinct, strong identity for this urban environment.

TEAM A: *Yasmin Naji from American University of Beirut (Lebanon), Yuka Kato from Meiji University (Japan), Carlos Santibanez from the University of Montreal (Canada), and Thea Sarkisian from the University of Montreal (Canada), under the faculty supervision of Iman Benkirane, National School of Architecture Rabat, Morocco, and Shao Yong, Tongji University, China.*

Figure 3.44: Public spaces around industrial artifacts.

Figure 3.45: Modifying the urban fabric and developing new architectural morphologies for the area.

Figure 3.46: Large vegetable interface of the road section.

ILLUSTRATION

SEEDS IN MONTREAL

The foundation of the *Seeds in Montreal* project rests upon a reading of the Montreal area that interprets the Cabot area as an island. This metaphor, closely related to the analysis of the geographical, historical, and cultural context of Montreal, reinforces the idea of a unique identity for the area.

The phrase "Think big, design small" guides the project design to analyze the problems of the site at the metropolitan and local levels and to use the studied area's intrinsic resources to solve them.

The suggested program is varied and meets social, cultural, sporting, recreational, and environmental vocations. It is progressive, reflects the seasons, adapts to opportunities and territorial constraints, and is developed using a long-term development strategy. Management solutions are simple and effective: create a productive identity park, multiply the Lachine Canal extensions, and develop a slope that allows views across the site. The great strength of this project is that it carries within it a story, a fantasy that feeds the aspirations regarding Montreal.

TEAM B: *Maha El Ayyoubi from the Lebanese University (Lebanon), Domenico Fogaroli from the University of Rome La Sapienza (Italy), Audrey Lavallée from the University of Montreal (Canada) and Naoko Yumoto from Kobe University (Japan), under the faculty supervision of Iman Benkirane, National School of Architecture of Rabat, Morocco, and Shao Yong, Tongji University, China.*

Figure 3.47: Planning the identity and productive urban park.

Figure 3.48: Conceptual diagram of the ecological use of the site.

Figure 3.49: Changing the topography and use of the site.

Site 6 / Downtown access corridor

MIND THE GAP

The stance of the *Mind the gap* project is to create a rallying linear park over the highway infrastructure located in the Ville-Marie area. The purpose of this intervention is mainly to engage in developing public spaces and innovative architectures that integrate highway and railway infrastructures in the process.

The urban design project counts on the opportunities of the adjacent neighborhoods to achieve its purposes, for example, by taking advantage of vacant spaces within the urban fabric, using sets of topography, and investing in buildings' roofs. This strategy also establishes the basis for a dialog between two historically and spatially splintered neighborhoods: Westmount and Saint-Henri. The area is to be developed in a way that meets the social, economic, and environmental context of the two neighborhoods.

Ultimately, this creates a "front" for the Saint-Henri district and reveals views of Montreal's urban landscape. This urban planning strategy allows both the enhancement and development of a disadvantaged neighborhood, as well as the manufacturing of a new landscape expression for the entrance in the heart of downtown Montreal.

TEAM A: *Emidio Arcidiacono from the University of Rome La Sapienza (Italy), S. Mohamed Ayari from the National School of Architecture and Urbanism (Tunisia), Juan Lin from Tongji University (China), and Valéry Simard from the University of Montreal (Canada)*, under the academic supervision of Sylvain Paquette, University of Montreal, Canada, and Julie Weltzien, American University of Beirut, Lebanon.

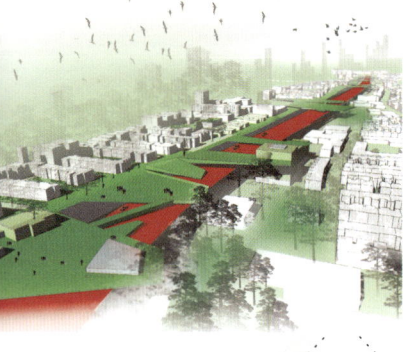

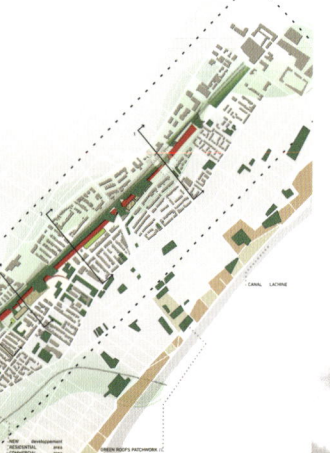

Figure 3.50: Aerial connection between Westmount and St. Henri neighborhoods.

Figure 3.51: Planning the linear park that overlooks Autoroute 20.

Figure 3.52: Green roofs act as a panoramic viewpoint near the Georges-Vanier metro station.

ILLUSTRATION

RE-COLLECTING MTL'S VITALITY

Figure 3.53: Urban densification of the western side of downtown and changing facades.

Figure 3.54: Planning the new renovated and improved Lucien-L'Allier station.

Figure 3.55: Planning the linear park that borders the highway and its transverse links to the canal.

The urban planning strategy of the *Re-collecting Mtl's vitality* proposal has two main purposes, namely to create a new downtown highway entrance experience and to establish lasting relationships between the highway infrastructure and the adjacent territories.

To do this, the urban design project first targets the development of green, cultural, and underground spaces along the highway infrastructure. The highlight of this corridor is a new building that houses the reinvented Lucien L'Allier station. Other interventions also improve downtown's quality of life, like greening both Guy and Atwater Street, renovating the facades, and concentrating buildings around the new station. The importance of this project, and what makes it unique, is that it induces ideas that more broadly benefit the entire city, and the invented solutions bring in a three-dimensional network of public spaces.

TEAM B: *Mazen Kandalaft from the University of Damascus (Syria), Marie-Pierre McDonald from the University of Montreal (Canada), Sara Saadoni from the National School of Architecture Rabat (Morocco), and Khuplianlam Tungnung from Kobe Design University (Japan), under the academic supervision of Sylvain Paquette, University of Montreal, Canada, and Julie Weltzien, American University of Beirut, Lebanon.*

Tools
Principles, Criteria And Planning Scenarios

4

The *YUL/MTL: Moving Landscapes* **international ideas** competition and the WAT_UNESCO Montreal 2011 allowed the identification of various strategies expressed from diagrams, plans, and illustrations that are inspiring for the future of the Autoroute 20 gateway corridor. The use of these images to engage in a constructive dialog for planning the territory proved to be a complex challenge to take. The success of these ideation exercises cannot only be measured by the quality of the ideas and by their communication, but also by the local stakeholders' assimilation of the ideas and their translation into specific urban planning measures. This last step of *YUL/MTL: Moving Landscapes* planning process is a challenge in itself and is based on a synthesis and communication of results. This chapter therefore aims to describe the synthesis process undertaken after the two ideation exercises and to present the action tools developed thereafter. These intervention tools are translated into design guidelines as well as planning scenarios.

Ideas competition, a strategic design process

The elaboration process of any spatial planning project, and particularly for highway infrastructure, is generally linear. That is to say that the creative phase associated with the design steps is integrated only after the planning phase is completed. In this linear process, the steps that follow the design are the drafting of plans and specifications and then the construction of the project. The ideas competition and the design workshops are based on a different relationship between design and planning, in which the creation of a particular design is directly integrated in the planning and collaboration process.[1] Here, the design is used to achieve better planning, to better identify the potential of a territory or a project, or even to illustrate the possibilities of implementing an urban planning vision as to facilitate its communication. In fact, "when they are within a shared vision development approach, workshops and design competitions represent the key steps in translating the aspirations and values shared by a community in terms of conservation, enhancement, and development of landscapes and

---

1. Van Dijk, Terry, 2011. Imagining future places: How designs co-constitute what is, and thus influence what will be, *Planning Theory*, 10(2): 124-143.

urban living environments."[2] To do this, the ideation exercises iteratively introduce the design within the planning process of a project. The result is a loop in which planning and collaboration are used as input to an ideation exercise whose results will improve the project planning to produce more relevant specifications in terms of the considered issues and more innovative ones in terms of the suggested solutions.

This distinction between linear and iterative process is important in terms of the interpretation of information that is expressed in the drawings, plans, and illustrations from the proposals originating from an ideas competition or a workshop. Thus, it is possible to distinguish the operational design, whose objective is to implement a project, from a strategic design that aims at drafting terms of reference.[3] A design with a strategic nature cannot be interpreted literally, by only considering how it is possible to implement a given proposal. In this sense, reading the drawings that come from an ideas competition or a workshop from an operational standpoint necessarily leads to the perception that the exercise's results are unrealistic, even utopian.

This impression is amplified at the territorial scale by the often-transversal character of the suggested visions that come from an ideation exercise, while on the field several administrative limits divide the area of intervention. The design strategies illustrated by the designers will necessarily go beyond these territorial limits to consider an overall portrait of the future. In addition, the designers rarely consider the complexity of the territory and the implementation of the suggested interventions. For example, an idea expressed in the context of an ideation exercise can overlap multiple jurisdictions and have different implementation temporalities. This gap between the images that result from a strategic design exercise and the implementation realities of a project shows the scale of the challenges to be taken by local stakeholders to develop a sense of ownership over the results. Local stakeholders may perceive the difficulty in considering their role in the implementation of a project, especially if the suggested actions exceed their power of action. Thus, in addition to the difficulty of fairly interpreting the design proposals, the many obstacles related to the capacity of the stakeholders to assimilate the results coming from ideation approaches reinforce the perceptions toward the unrealistic nature of the results of an ideas competition.[4]

---

2. Poullaouec-Gonidec, P. and S. Paquette, 2011. *Montréal en paysage*, Presses de l'Université de Montréal: Montréal, p. 222.
3. De Jonge, J., 2009, *Landscape architecture between politics and science. An integrative perspective on landscape planning and design in the network society*. PhD thesis Wageningen University. P.17.
4. Rauws, W. and T. Van Dijk, 2013. A design approach to forge visions that amplify paths of peri-urban development, in *Environment and Planning B : Planning and Design*, 40 : 254-270.

The success of strategic ideation exercises is measured through local appropriation of ideas and their translation into concrete urban planning framework.

To demonstrate the full contribution of a prospective design exercise within a planning process, it is therefore necessary to reveal its strategic nature and scope. During the Autoroute 20 gateway corridor project, the results synthesis quickly appears to be essential. The concept of an "atlas of possibilities" presented by the jury of the *YUL/MTL: Moving Landscapes* ideas competition illustrated the strategic reach of the exercise well. Far from putting forward one single proposal by awarding one prize, the jury appointed three tied winners and accompanied these nominations by a range of recognitions that complemented the winning proposal's contribution. If the selection of winners helped to emphasize the qualities of three overall strategies for the future of the territory and of the infrastructure, the identification of specific ideas from a dozen additional proposals offered very useful additional information to guide the sequence of the planning process. Besides identifying the most interesting proposals in regards to the program and ambitions of the competition, this decision emphasized the interest for all received proposals, which made a pool of ideas to enrich a reflection on the future of the Autoroute 20 gateway corridor.

The results synthesis from the *YUL/MTL: Moving Landscapes* project, based on its two ideation exercises (ideas competition and WAT_UNESCO), is inspired by the "atlas of possibilities" idea launched by the jury to stimulate local collaboration between the territorial stakeholder members of the committee of the Autoroute 20 gateway corridor. However, to our knowledge, few comparable precedents were implemented to date. At least two processes, however, merit particular attention in this sense. On the one hand, the international workshops of the Grand Paris[5], a broad collaboration launched in France in 2008 by

---

5. http://www.ateliergrandparis.fr/

the ministère de la Culture et des Communications, which is probably the greatest effort of forward-looking design of the last decade. The planning visions that were imagined by 10 teams of internationally renowned architects, landscape architects, and urban planners, as well as the reflection work that followed, helped to define the metropolitan territory of Paris – to identify its contours and characteristics and to consider the needs in term of territorial planning. On the other hand, the international ideas competition on the 2050 Vision for the Greater Helsinki[6] first aimed at creating a mental picture of the metropolitan area limits of Helsinki and its possibilities. Similar to the *YUL/MTL: Moving Landscapes* international ideas competition, the jury of the competition on the Greater Helsinki did not hand out one unique prize, but rather it identified nine winners stressing the complementarity of the proposals. Following the Greater Helsinki ideas competition, the synthesis and collaborative process spread over a period of one year and allowed the identification of a series of themes and key ideas to animate a regional collaboration and articulate the strategic planning vision of the metropolitan territory.

More recently, two large operations of prospective design, "Changing Course"[7] and "Rebuild by Design"[8], were launched in the United States to rethink the development of vast regions affected by hurricanes Katrina (Louisiana) and Sandy (New Jersey/New York). Put forward with the participation of the Van Alen Institute, these exercises are still too recent to assess their real contributions. They nevertheless evoke a growing interest to use design in a strategic manner, and illustrate the role that designers can play in the conceptualization of the territories as well as the tools to be developed to achieve this.

These recent examples thus testify for the increasing need to intervene on large scales to plan the territory and the difficulty of using the usual regulatory mechanisms to act on these scales. The multiplicity of stakeholders and interests makes finding a consensus difficult. It is in this context that "soft planning" tools are appearing to inspire action rather than to constrain it. The ideation exercises are part of these efforts which foster the sharing of visions in order to inspire the future.[9]

---

6. Ache, P., 2011. « Creating futures that would otherwise not be » - Reflections on the Greater Helsinki Vision process and the making of the metropolitans regions, *Progress in Planning*, 75 : 155-192.
7. Changing Course, http://changingcourse.us/
8. Rebuild by Design, http://www.rebuildbydesign.org/
9. Allmendinger, P. and G. Haughton, 2009. Soft spaces, fuzzy boundaries, and metagovernance : the new spatial planning in the Thames Gateway, in *Environment and Planning A*, 41 : 617-633.

Out of 73 submitted proposals, 500 ideas were listed.

YUL/MTL ideas competition and WAT_UNESCO Montreal: analysis approach and proposals overview

An in-depth analysis of all proposals designed during the *YUL/MTL: Moving Landscapes* international ideas competition and the WAT-UNESCO Montréal was carried out. More than a simple activity report, this results synthesis made the establishment of the necessary foundation for the continuation of the collaborative planning process possible, and most importantly, the development of common tools to plan the Autoroute 20 gateway corridor.

Although taking the form of overall strategy, each of the proposals contains a large number of individual intervention ideas. Some ideas directly concern the road infrastructure; others are more oriented toward the transformation of the territory adjacent to the highway. To facilitate the sense of ownership towards ideas, it was necessary, as a first analysis step, to perform a deconstruction work in order to obtain a fair appraisal of the different types of intervention suggested for the different concerned areas or stakeholders. Thus, a careful reading of the proposals coming from the YUL/MTL ideas competition and the WAT-UNESCO Montréal was undertaken in order to list all the individual ideas issued. Each individual idea had to correspond to an action that can be independently undertaken compared to other actions recommended in the overall strategy. Similarly, this action was meant to present the implementation potential by a single territorial stakeholder or through the coordination of a limited number of stakeholders.

In total, from 73 proposals designed during the two ideation exercises, 500 ideas were listed. For each of them, the area of implementation was identified as well as the intentions of the designers. Certainly, several of these individual ideas showed similar characteristics. Thus, by grouping individual ideas, it was possible to identify 47 intervention strategies suggested by the designers. They form the heart of the synthesis, because they summarize the proposals. They also constitute a set of easily appropriable actions that were classified according to their relationship to the highway right-of-way or to the neighboring living environments. In so doing, it also facilitated the recognition of the stakeholders' responsibilities in the implementation of the generated ideas.

Based on the designs intentions communicated in the proposals through images, texts, annotations, and videos, each intervention strategy was described to explain the implementation principles that responded to the initial vision of the competition. It was thus possible, beyond the guidelines listed in the collective vision and written ahead of the competitions, to identify strategies detailed through design guidelines. This benefit of the approach thus constitutes a genuine operational methodology for the planning of specific projects.

If these guidelines are essential in this sense, the overall result remains too large however to facilitate and enrich the collaborative process. Starting from the "atlas of possibilities" expressed in the international competition and the design workshop, a sorting, ranking, and prioritization step seemed unavoidable and even necessary. That is why the strategies of interventions were grouped in the form of planning scenarios from reading the themes that appear as convergent with respect to intervention. The five developed scenarios offer alternative visions of a possible future such as imagined by the original designers of the YUL/MTL international ideas competition and the WAT_UNESCO Montreal.

This sequence of deconstructing / reconstructing the designers' proposals aims to promote the results of the ideation exercises to policy makers. Following the presentation of this synthesis exercise, the positive reaction of the Autoroute 20 gateway corridor work table quickly demonstrated an increased understanding of the designers' proposals and their potential contribution to planning over a longer term. In doing so, this simplification of the results, made possible thanks to the development of scenarios, will greatly facilitate the dialog between the stakeholders of the working group and debates about the major intervention strategies in order to prioritize the future of the city entrance corridor.

The following lines outline the results of this synthesis with an emphasis, first, on the main intervention strategies that emerged. Second, the five planning scenarios for the Autoroute 20 gateway corridor will be addressed.

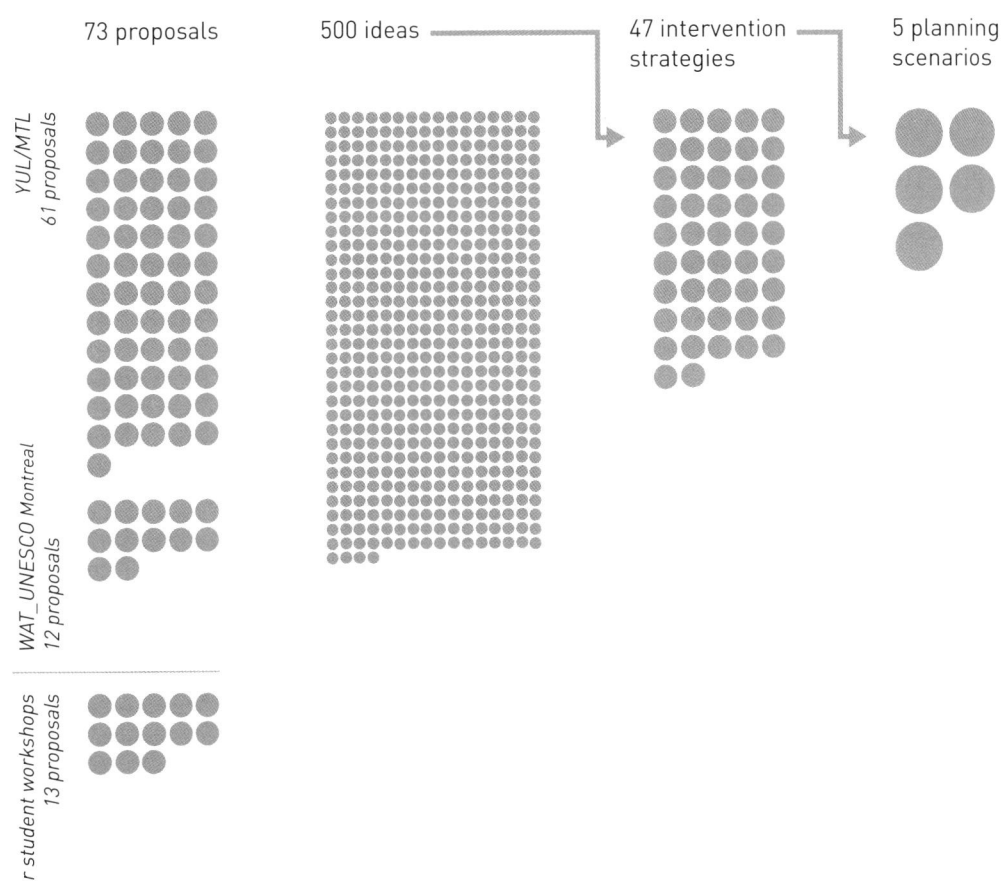

Figure 4.1: Reading and analysis of proposal process.

Intervention Strategies

The premise behind the collaborative approach undertaken by the Autoroute 20 gateway corridor working group is that each individual intervention of the local stakeholders, although carried out separately from each other, sways the development of the territory and its landscape expressions. Thus, collaboration allows for a maximal coherence of the planning and urban projects processes by the coordination of individual actions.

From this premise, two essential questions guided the identification and grouping of intervention strategies:

- How to emphasize the landscape interactions that exist between the interventions carried out by the different territorial stakeholders?
- How to identify the types of landscape interaction that are present in the proposals designed during the YUL/MTL international ideas competition and the WAT_UNESCO Montreal?

Thus, from compiling the intervention strategies, a clear division emerged between those directly affecting the highway development and its right-of-way and those that aimed at the management of adjacent environments or other transportation infrastructures of the city entrance corridor.

Although this distinction allows identifying the stakeholders responsible for the implementation of an idea, it does not emphasize, however, the visual relationships at work in the territory of the city entrance corridor. A more careful review of the intervention strategies shows that a very large proportion demonstrates the complex and dynamic relationships that exist between the transportation infrastructure and its territory. Thus, several interventions that aim to redefine a composition element of the highway aim to enhance the territory and the adjacent environments. Conversely, several interventions that redefine the adjacent environments are based on the will to enhance the highway landscape.

From this observation, five categories of intervention strategies were established to present information and to continue the analysis. These categories are:

1. The highway strategies that are directly designed to redefine the highway landscape. The relations to the territories, although possible, are generally marginal.

2. The strategies in which the highway acts as an aesthetic and visual element of the territory. They modify the formal appearance of the highway components in order to emphasize the qualities of the local landscape or promote a positive perception of the highway from the adjacent neighborhood.

3. The strategies in which the highway acts as function-enhancing element of the territory. These are designed, for example, to facilitate the movement (all methods of transport combined), the neighborhood's interconnectivity, or the development of adjacent environments.

4. The strategies in which the territory acts as an enhancing element of the highway experience. These strategies are applied to the neighborhoods adjacent to the highway in order to improve the landscape experience from the highway.

5. The strategies that are intrinsic to the territory. They aim to requalify the adjacent living environments and have little impact on the landscape experience from the highway. These strategies often involve the relationship between the living environments, the natural environments, and other transportation infrastructures, such as rail tracks or the Lachine Canal.

The following sub-section presents the various intervention strategies that come from these categories. In addition to a brief description of the strategy, each page outlines the main intentions, the design criteria, and a few illustrations that are associated with it.

CATEGORY 1
The highway intrinsic strategies

Figure 4.2: Selection of tree species according to their color (Excerpt from the proposal made by Gerwin de Vries + Alexander Herrebout, The Netherlands)

Figure 4.3: Planting trees in the highway right-of-way (Excerpt from the Catalyse urbaine proposal, Canada)

1.1

REVEGETATION OF HIGHWAYS

While aiming at getting nature/infrastructure closer through the revegetation of the roadbeds, these interventions also aim at defining interfaces between the highway and the adjacent environments. The vegetation is used to filter the nuisances caused by the highway as well as to create friendly spaces at the borders of the highway right-of-way.

The vegetation also allows perceptual changes to the highway landscape both at the ground level and on elevated structures to be made. On the ground, the perception of the highway landscape continuity is improved by viewing the trees in alignment. On the elevated structures, users can see the new urban forest canopy like a hanging garden. The perception of Montreal from the highway is that of a city in which vegetation is present, thus putting forward an image of excellence for sustainable development.

The vegetation also allows for a better perception of the identity of Montreal as a Nordic city, especially by the effect of seasonal changes of the plant's color.

MAIN IDEAS

- Development of vegetated spaces in the residual interstices of the highway right-of-way
- Plant systems integrated to highway structures
- Creation of seasonal ambiances by the selection of species according to their color

CRITERIA FOR DESIGN

- Promote the perception of seasonal changes through vegetation
- Enhance the sustainable development image of Montreal through a greater presence of plants
- Fasten the revegetation strategies to the existing natural areas (Saint-Jacques Escarpment, residual vegetation of the Dorval and Lachine areas) in order to contribute to their enhancement
- Design a planted interface between the highway and the adjacent environments
- Develop the residual spaces of the highway right-of-way through the creation of green spaces
- Plan an ecological corridor to ensure the habitats continuity
- Filter nuisances coming from the highway

1.2
NOISE-REDUCING WALLS

With the expressive character of the highway journey, the installation of artworks, and even the construction of a highway space (ex.: acoustic tube), allows the creation of new highway experiences.

The noise-reducing wall generates a dual landscape experience, that of the highway and the adjacent living environments. The differential treatment of these two sides of the noise-reducing wall allows the creation of coherent interfaces with the present uses. The presence of vegetation or artwork on the inhabited side of the noise-reducing walls facilitates the development of friendly public spaces.

Some proposals explore the multifunctional character of the walls through the possibility of integrating them to the public space or even giving them an energy production function.

MAIN IDEAS
- Treatment of the interface between the highway and the adjacent environments through the construction of a noise-reducing wall
- Creating ambiances through the artistic treatment of the wall

CRITERIA FOR DESIGN
- Make the noise-reducing walls represent the creative identity of Montreal (e.g.: marking and communication of public interest)
- Enhance the highway structures
- Use new technologies to develop the multifunctionality of the noise-reducing walls (e.g.: energy production, signage)
- Consider the double relationship of the noise-reducing walls, road corridor, and living environment
- Facilitate the appropriation of public spaces bordering the noise-reducing walls through the presence of plants or artworks
- Develop planting systems for the noise-reducing walls
- Filter nuisances coming from the highway

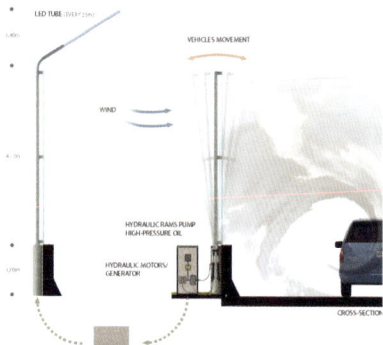

Figure 4.4: Revitalization of the noise-reducing walls (Excerpt from a proposal of the WAT_UNESCO Montreal workshop by: Bruna Bajramovic, Paolo Casu, Stephany Khoriaty, Amayel Ndiaye)

Figure 4.5: Production of energy using wall vibrations (Excerpt from the proposal of PRT PLAN, Portugal)

1.3

TREATMENT OF RUN-OFF WATERS

Run-off water collection fits into an environmental perspective as much to facilitate the natural filtration of the water as to avoid the drainage equipment overload (storm sewer flooding). Run-off water collection is also considered for its potential to transform the landscape by promoting the presence of a richer flora and fauna.

This intervention strategy also aims to evoke the specificity of Montreal as an island. Basins allow the construction of bridges with their feet in the water evoking a river crossing. They also offer the possibility of developing new uses related to the water such as fish farming.

MAIN IDEAS
- Development of wetlands on the edge of the highway for the collection and treatment of run-off water

CRITERIA FOR DESIGN
- Use run-off water and create basins to renew the territorial identity of corridor
- Recall Montreal's island identity through watershed management under the elevated structural pillars (e.g.: metaphor and narrative)
- Promote new water treatment technology (e.g.: phytoremediation - storm water management system)
- Facilitate the appropriation of adjacent living areas by using water as an enhancement feature
- Reduce the environmental impact from run-off water coming from the highway (e.g.: phytoremediation - harnessing of heavy particles)
- Use water to create distinctive landscape ambiances (e.g.: sounds from waterfalls or fountains)
- Promote the presence of diverse ecological habitats
- Use infrastructure elements (system of storm-water management) to enhance the territory adjacent to the highway

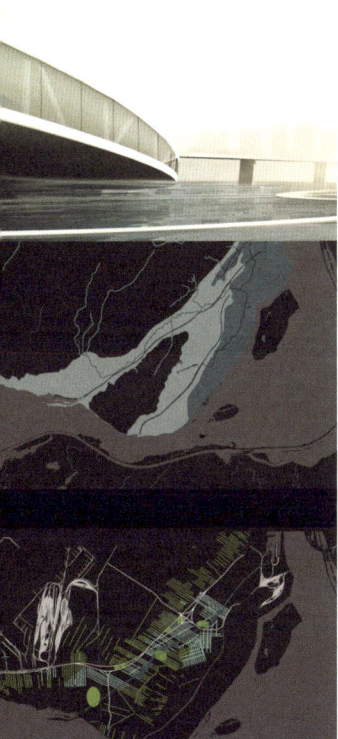

Figure 4.6: Lake collecting rainwater (Excerpt from the proposal made by ZerOgroup + Fabrica de paisaje, Brazil)

Figure 4.7: Development of wetlands on the edge of the Lachine Canal. (Excerpt from the proposal of dlandstudio, United States)

1.4

MAINTENANCE AND RENOVATION OF EXISTING STRUCTURES

These interventions start from the standpoint that the deteriorated appearance of infrastructure projects a negative image. They aim to renovate the structures to counter this impression. It is suggested that these interventions be conducted early in the process of territorial requalification.

MAIN IDEAS
- Repair of highway structures, including the interchanges and the Ville-Marie tunnel, to give them a new look

CRITERIA FOR DESIGN
- Enhance the sustainable development image of Montreal through the repair of existing structures
- Enhance the highway structures by the quality of their physical condition and their cladding materials
- Promote the quality and sustainability of civil engineering structures
- Strengthen the impression of security through the maintenance quality of the civil engineering structures

Figure 4.8: Renewal of cladding materials and lighting of the Ville-Marie Tunnel. (Excerpt from the proposal of Team KDU, Japan)

CATEGORY 2
The highway strategies for the aesthetic and visual enhancement of the territory

2.1
COLOR OR LIGHT INTERVENTIONS ON THE HIGHWAY STRUCTURES

Often denigrated for their brutal character, the highway structures are enhanced with creative and artistic approaches. These interventions are thus used to emphasize the qualities of the infrastructure as civil engineering structures, or even to be aware of the potential heritage value of the infrastructure. They also allow new identities to be given by creating staging and luminous expressions.

These interventions enhance specific locations (the interchange) and thus stress the highlights of the highway journey. They also give to the interchanges land-marking value at night for the user and the residents who can develop a positive perception of the infrastructure.

MAIN IDEAS
- Enhancement of the architectural qualities of the highway structures (pillars and concrete slabs, noise-reducing walls, pavements) by the use of a coloring or lighting scheme
- Creation of different day/night ambiances

CRITERIA FOR DESIGN
- Highlight the identity of Montreal as a creative city
- Enhance the highway structures by emphasizing their aesthetic qualities
- Use the new lighting technology to design nocturnal ambiances
- Mark the pathway progression by creating visual landmarks
- Ensure the consistency of nocturnal ambiances. The lighting and coloring scheme can be used to coherently structure these landscape ambiances

Figure 4.9: Enhancement through light. (Excerpt from the proposal of Johnathan Yue, Singapore)

Figure 4.10: Coloring the highway's structures. (Excerpt from the proposal of Catalyse urbaine, Canada)

2.2

LIGHTING EQUIPMENT

The interventions on the lighting equipment seek to create ambiances that evoke the local vegetation or that aim to show the level of activity of the city through a variation of light intensity. Elements difficult to perceive at night are thus brought forward in the nocturnal landscape of the highway.

The creation of lighting equipment also allows the insertion of strong identity markers in the daytime landscape thus playing with the perceptions of day and night.

MAIN IDEAS
- Use of urban equipment, including lighting systems, to create ambiances in the nocturnal landscape of the entrance pathway
- The interventions use as much functional lighting for the highway as ambiant lighting

CRITERIA FOR DESIGN
- Highlight the identity of Montreal as a creative city (e.g.: visual landmark of public interest)
- Creation of daytime and nighttime ambiances by developing innovative lighting concepts and using new technologies
- Designing lighting equipment that ensures both road safety and the creation of distinctive styles (e.g.: lighting schemes)

Figure 4.11: Lighting trees. (Excerpt from the proposal of Gilles Hanicot, Canada)

Figure 4.12: Red frames, carriers of light. (Excerpt from the proposal of MAP+OPO, Canada)

Figure 4.13: Art installation along the highway. (Excerpt from the proposal of Andrew Forster, Canada)

Figure 4.14: Framework emphasizing the Montreal landscape elements. (Excerpt from the proposal of Thibodeau Architecture + Design, Canada)

2.3

INSTALLATION OF ONE OR SEVERAL EMBLEMATIC ARTWORKS

The main objective of these interventions is to highlight the creative identity of Montreal through the presence of artworks on the pathway, like the Eastlink Highway in Melbourne. A similar idea was developed in 1976 for the Sherbrooke Street in Montreal (Corrid'art).

The artworks create ambiances on some sections and serve as a landmark. The *Paysages suspendus* competition in Quebec City or the intervention at the airport entrance of Los Angeles (LAX) constitute similar examples of marking specific places in the highway landscape.

Some of the suggested artworks were developed through an investigation of identity archetypes for Montreal or Canada (beaver, inuksuk, industrial container, aerospace industry). Others stress the peculiar beauties of the landscape or aim at creating new landmarks.

Using ground coverage to create an artwork visible from the air is one of the rare considerations for the entrance pathway for the air transportation users.

MAIN IDEAS

- Creating artworks that stage the landscape of the city entrance pathway
- The artworks are positioned on strategic points of the corridor (Dorval Interchange, Turcot Interchange) or are spread over the entire pathway

CRITERIA FOR DESIGN

- Highlight the identity of Montreal as a creative city (e.g.: artistic marking regarding the industrial past of the transportation corridor)
- Enhance landscape expressions through the artworks
- Use the industrial theme (containers) as a source of inspiration
- Mark the pathway progression by creating soundscapes or visual landmarks
- Contribute to the urban animation of residual spaces of the highway right-of-way

2.4

INSTALLATION OF LANDMARKS AT THE NEIGHBORHOOD ENTRANCES

This intervention strategy emphasizes the identity of Montreal as a sum of local identities anchored at the neighborhood level. Thus, with the presence of markers at the entrance of neighborhoods, the highway user understands this territorial structure. These strategies fit into the logic of legibility and intelligibility principles of territorial structures developed by Kevin Lynch. In this theory, the presence of landmarks allows to better locate the attractions and therefore to anticipate future decisions taken along the pathway.

The architectural expression of the landmarks can be anchored in the local heritage (industrial structure, clock tower) or the creation of new ambiances.

The development of bridge structures concomitantly with the creation of landmarks allows the creation of entrances in unique neighborhoods for several courses (highway exits, railway stations, cycling lanes).

MAIN IDEAS
- Intervention strategy on the whole city entrance corridor that aims at marking the neighborhood entrances by the construction of a landmark
- These objects can incorporate a display component, serve as a bridge or be inhabited

CRITERIA FOR DESIGN
- Reveal the identities of the local territories
- Mark the pathway progression by creating visual landmarks
- Facilitate the urban environment legibility through the presence of a strong exit landmark

Figure 4.15: Tower signs at the exits of the highway. (Excerpt from the proposal of PRT PLAN, Portugal)

Figure 4.16: Observation tower and gateway in Lachine. (Excerpt from the proposal of Catalyse urbaine, Canada)

2.5
SOUND/VIDEO INSTALLATIONS

These intervention strategies are designed to create ambiances that evoke Montreal's identity through sensory exploration. They allow a user to become aware of the nature of the local culture through historical, literary, or musical sound clips. The use of new technologies (e.g.: georeferenced podcasting) allows users to receive information during their trip in an audio guide manner.

The sound installations may also be considered in parks and urban public spaces to create ambiances that filter the highway noise. A similar experiment was conducted in Montreal during the Décarie highway reconstruction to reduce the noise coming from the construction site at night.

Sound installations are also considered for their ephemeral nature. Thus, the experiences created may vary frequently to give a renewed experience to users who regularly use the highway. They also allow specific information to be provided at appropriate times, such as during construction work.

MAIN IDEAS
- Sound or visual installations that can be captured from multiple technology platforms

CRITERIA FOR DESIGN
- Highlight the identity of Montreal as a creative city through the use of audio ambiances
- Giving information of public interest about places Using new information technologies (e.g.: podcasting)
- Marking the pathway progression by creating audio ambiances
- Filter noises coming from the highway by using audio ambiances

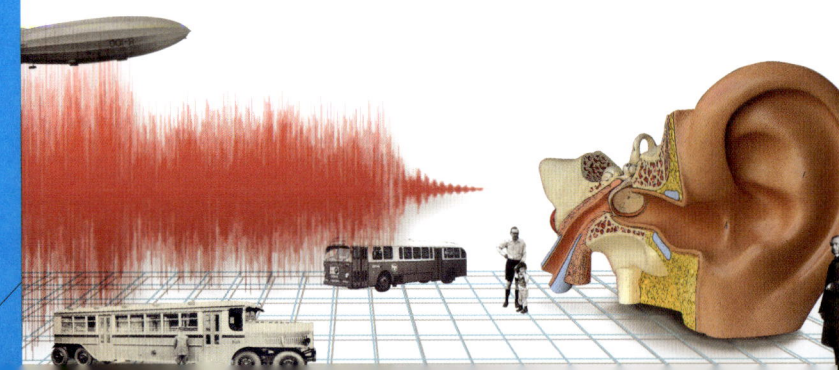

Figure 4.17: Audio perception of the pathway characters. (Excerpt from the proposal made by Andrew Forster, Push Montreal, Canada)

CATEGORY 3
The highway strategies for the functional enhancement of the territory

3.1

CONSTRUCTION OF BRIDGE/FLYOVER TO CROSS THE HIGHWAY

The perception of the highway and other transportation infrastructures as a barrier limiting active transportation is present in a large number of proposals. The solutions suggested to address this barrier issue through a range of possibilities, from the simple gateway to inhabited bridges and ecological bridges (e.g.: strategies for ecological bridge were also explored in the context of the ARC - Wildlife crossing infrastructure design competition). These last two strategies aim to not only facilitate travel, but also promote the vitality and sustainability of the territory to create animated places.

The bridge is also a strong marking element of the territory. Through research on their architectural expression or their land-marking value, bridges allow the creation of new territorial identities.

MAIN IDEAS
- Planning pedestrian bridges to facilitate crossing the highway and accessing strategic locations of the city entrance corridor (e.g.: Lachine Canal, Saint-Jacques Escarpment, railway stations)
- Development of ecological bridges allowing the continuity of natural environments

CRITERIA FOR DESIGN
- Investing in the creation of unique and emblematic infrastructures (project competition)
- Develop public spaces integrated into the civil engineering structures
- Mark the pathway progression by creating visual landmarks
- Development of alternative pathways to the highway (cyclists, pedestrians)
- Develop attractive new public spaces
- Integrate the bridges design to the establishment of natural and recreational corridors
- Facilitate crossing the highway while ensuring territorial continuities (uses, functions, etc.)

Figure 4.18: Inhabited bridge whose wall allows signage. (Excerpt from the proposal of Ghazal Jafari + Ali Fard, Canada)

Figure 4.19: Emblematic gateway (TATINVESTGRAZG-DANPROJECT, Russia)

3.2

CONSTRUCTION OF A MULTIFUNCTIONAL TRAIL NETWORK (HIGHWAY)

The creation of multifunctional trails is considered to promote active transportation. Often planned for cyclists or for the promenade, some interventions also consider cross-country skiing or horse riding as possible use of these trails.

The implementation of a multifunctional trail inside the highway right-of-way allows for an improved connection to the existing bicycle network in the neighborhoods. Thus, the network created with the highway right-of-way contributes to the development of the local network.

The experience of the multifunctional trail is seen as a pathway in itself, for which the variation between day and night ambiances must be explored.

The trail facilities and the presence of users also animate the corridor urban space thus following the logic of renewing the territory vitality.

MAIN IDEAS
- Development of multifunctional trails in residual spaces of the highway right-of-way and anchoring them with the existing network of bicycle lanes

CRITERIA FOR DESIGN
- Enhance the sustainable development image of Montreal
- Develop alternative pathways to the highway (cyclist, pedestrian-sustainable mobility plan)
- Contribute to the urban animation and enhancement of residual spaces of the highway right-of-way
- Integrate the design of multifunctional trails to the establishment of natural and recreational corridors

Figure 4.20: Multifunctional tracks along the Lachine Canal. (Excerpt from the proposal of David Garcia Studio, Denmark)

Figure 4.21: Multifunctional trails under the Turcot Interchange. (Excerpt from the proposal of Ghazal Jafari and Ali Fard, Canada)

3.3

COMPLETE OR PARTIAL BURYING OF THE HIGHWAY

Burying the highway is generally not considered a part of the perspective to create particular driving experiences, but rather to ensure the continuity of the urban networks and suggested urban developments. Thus, this intervention strategy aims to offer a response to the barrier effect created by the highway. The former Turcot rail yard is a privileged place for the construction of tunnels, because it is framed by two remarkable elements of the city entrance corridor, the Saint-Jacques Escarpment and the Lachine Canal. This intervention strategy fits into the same logic as the burying of Boston's Highway 93 and the creation of parks in the freed right-of-way.

Nevertheless, a few interventions do emphasize the potential to create specific highway experiences including the discovery of a new landscape at the tunnel exit. The latter may eventually result in a surprise effect because the downtown skyline would be revealed.

MAIN IDEAS
- Burying some sections of the highway particularly in the former Turcot rail yard to allow a continuity of urban developments at ground level

CRITERIA FOR DESIGN
- Promote new public spaces that participate in the development of the territory
- Develop alternative pathways to the highway (cyclist, pedestrians) and public spaces
- Generate visual landmarks by creating surprise effects at the exit of the tunnels
- Develop territorial and landscape continuities
- Encourage planning and development of the neighborhood adjacent to the highway
- Create new opportunities for revegetation
- Improve connections between neighborhoods by reducing barriers
- Reduce nuisances coming from the highway

Figure 4.22: Burying Autoroute 20 in the former Turcot rail yard to facilitate development. (Excerpt from the proposal of dlandstudio, United States)

Figure 4.23: Partial covering of Autoroute 20. (Excerpt from the proposal of dlandstudio, United States)

3.4

CREATING PUBLIC SPACES IN RESIDUAL SPACES OF THE HIGHWAY RIGHT-OF-WAY

Free spaces under the interchanges are generally undervalued places. These interventions are designed to give them a strong urban nature that facilitates their appropriation. Some interventions mainly rely on urban planning that promotes only the temporary occupation or the pathway (multifunctional lanes) or space occupation by target groups (skate park), but other proposals are aimed at a more sustained occupation through the installation of kiosks or a more important revegetation of the public space.

These interventions also fit in a revaluation of urban highway structures in a manner that is similar to the heritage enhancement approaches put forward by other intervention categories, particularly lightning.

MAIN IDEAS
- Urban planning of public place in residual spaces from the highway right-of-way, particularly the free spaces under the structure of the interchanges
- Animation of these public places through the installation of recreational equipment

CRITERIA FOR DESIGN
- Develop the highway structures by promoting social appropriation of their urban environment
- Create infrastructures that are more suited to receive public uses (e.g.: passages)
- Contribute to the urban animation and enhancement of the residual spaces of the highway right-of-way
- Valuing the revegetation of infrastructure/territory interfaces
- Create the interfaces that minimize the highway infrastructure and value the territorial ownership

Figure 4.24: Public space under the Turcot Interchange (K2T2R, Singapore)

Figure 4.25: Creating a public park under the Turcot Interchange (Excerpt from the proposal of Ghazal Jafari and Ali Fard, Canada)

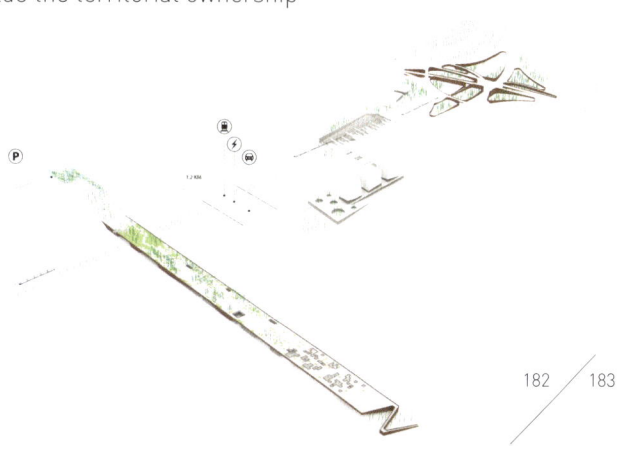

3.5
MEGASTRUCTURE

These intervention strategies fit into a perspective of strong integration of the territorial stakeholders to achieve a multifunctional infrastructure project. The networks superposition highlights the possible interconnections between the methods of transportation and the potential for diversity and densification of uses near the nodes (stations, highway exits, and terminus).

The megastructures also facilitate pedestrians' movement by integrating gateways or platform parks above the highway.

MAIN IDEAS
- Construction of an infrastructure at multiple levels that integrates several types of transportation (highway, railway, cycling lane) or urban uses (parking, public buildings)

CRITERIA FOR DESIGN
- Pair the infrastructures to create strong and emblematic elements in the city entrance corridor landscape
- Create new uses and territorial functions (e.g.: poles of development)
- Create common nodes for several methods of transportation to generate visual landmarks
- Multiply transportation opportunities in the highway right-of-way
- Maximize the economic and social value of the targeted territories
- Ensure a better integration of the highway infrastructure with the urban development using multimodal pole

Figure 4.26: Integration of the Dorval Interchange (Excerpt from the proposal of the workshop WAT_UNESCO Montreal by: Andrea Becca, David Fiset, Natsuko Kobayashi, Zeineb M'Rabet)

Figure 4.27: Platform park above the highway (Excerpt from the collective proposal of Philippe Barrière, Canada)

3.6
SIGNAGE

The intervention strategies that suggested sign and display elements sought to make local landmarks legible, thus placing emphasis on the singularity of places that make up the city entrance pathway and revealing the identity of Montreal as a creative or sustainable city.

The signage systems also transform the corridor in an informative pathway where users are informed about the weather or local events in addition to road conditions.

The use of a common signage system for different entrance paths (highway, public transportation, bicycle lanes) also allows for the interaction and complementarity of the different methods of transportation. The nodes and interconnections of transportation networks are highlighted by the identification of terminals or transit hubs.

MAIN IDEAS
- Development of a visual identity for the city entrance corridor. This visual identity may be common to all routes and identifies the main nodes and points of interest
- Enhancement of Montreal through the development of display systems for walls and highway equipment

CRITERIA FOR DESIGN
- Develop a signage identity in addition to the regular signage
- Develop new systems for the marking of public interest information Mark the pathway progression by creating visual landmarks
- Design a common visual identity for all the city entrance corridor pathways
- Stimulate the attractiveness of the local territories
- Build on the multifunctionality of the transportation infrastructures

Figure 4.28: Highlighting points of interest through signage (Excerpt from the proposal of David Garcia Studio, Denmark)

Figure 4.29: Banner with logo on the infrastructure (Excerpt from the proposal of Gilles Hanicot, Canada)

3.7

IMPLEMENTATION OF ENERGY PRODUCTION EQUIPMENT ALONG HIGHWAY INFRASTRUCTURES

In these intervention strategies, the highway equipment takes advantage of different renewable energy sources (wind, solar, geothermal) to supply electricity or heat to the road network and adjacent environments.

While some of the strategies are aimed at the establishment of innovative systems (ex.: noise-reducing wall reacting to car movements), others instead take advantage of the opportunity created by the interchange reconstruction site to install energy production systems for which technology exists today (geothermal system concomitantly installed to the excavation of the foundations of the future interchange pillars).

The energy strategies target a double identity change of the highway corridor. On the one hand, they aim at transforming the perception of an energy consumer system toward an energy producer system thus emphasizing the sustainable nature of the infrastructure. On the other hand, they allow for the creation of new ambiances along the corridor, thus contributing to its landscape staging.

MAIN IDEAS
- Using highway equipment to produce energy used for the road network (lighting) or redistributed to the adjacent environments

CRITERIA FOR DESIGN
- Highlight the sustainable development image of Montreal
- Develop new technologies to implement the infrastructure multifunctionality including energy production
- Creation of visual landmarks
- Facilitate snow removal and ice control by using the energy produced to heat the ground
- Transform the transportation corridor in a place of energy production
- Supply the adjacent environments with energy (heating, electricity)

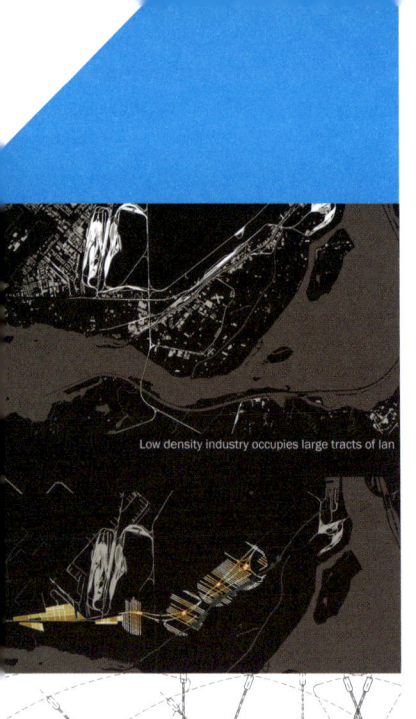

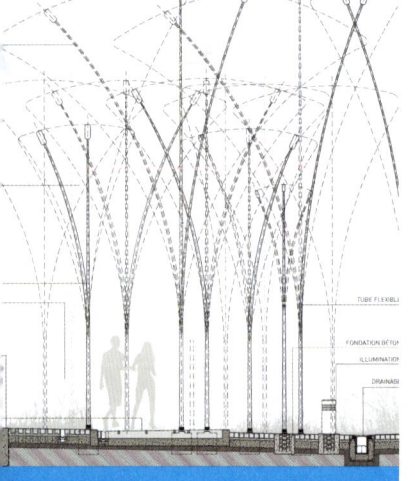

Figure 4.30: Distribution radius for geothermal energy (Excerpt from the proposal of dlandstudio, United States)

Figure 4.31: Piezoelectric production structure (Excerpt from the proposal made by ZerOgroup + Fabrica de paisaje, Brazil)

3.8

CONSERVATION OF THE TURCOT INTERCHANGE RAMPS

Two distinct motivations underlie the conservation of obsolete structures of the Turcot Interchange: on the one hand, this structure has a strong potential for marking the pathway progression to downtown; on the other hand, its development would contribute to the enhancement of the heritage value of major construction works from the 1960s. This last approach emphasizes the growing trend towards the recognition of the heritage value of modern architecture and civil engineering structures. Combined, these two approaches to enhance the current Turcot interchanges structures stress the place's identity with the artifacts from which it was firstly built.

Beyond the identity strategies, interventions for the development of the Turcot Interchange current structures also propose their rehabilitation in a project of aerial pedestrians' walkways. These strategies are founded on the multiple proposals for the development of bridges or multifunctional lanes to preserve the experience of exceptional visual perspectives toward downtown and the region, which are possible from the top of the current interchange. They are inspired by examples of projects such as the High Line in New York, where elevated rail tracks were transformed into a linear park.

MAIN IDEAS
- The obsolete structures of the former interchange (pillars and girts) are retained and reused in the new development project of the Turcot complex

CRITERIA FOR DESIGN
- Highlight the sustainable development image of Montreal
- Develop the highway structures of the Turcot Interchange as heritage components
- Create infrastructure recycling projects for public uses (e.g.: linear park, aerial promenade)
- Mark the pathway progression by creating visual landmarks, staging, urban belvederes
- Offer a new public space for urban walks
- Develop new territorial links by reusing old ramps

Figure 4.32: Reusing outdated ramps of the Turcot Interchange for the development of elevated multifunctional lanes (Excerpt from the proposal of Catalyse urbaine, Canada)

3.9

HIGHWAY REALIGNMENT

The interventions that are considering the realignment of the highway seek to create new relations in the territory of the city entrance corridor. Thus, this operation is mainly considered to facilitate the interconnection of the neighborhoods and the reduction of the highway right-of-way. Although these interventions modify the landscape experiences seen from the highway, their objective is mainly to improve the quality of life and facilitate the development of the territory.

MAIN IDEAS
- Realignment of certain Autoroute 20 sections

CRITERIA FOR DESIGN
- Decrease the infrastructure footprint to promote the development of adjacent environments
- Improve connections between neighborhoods by reducing barriers

Figure 4.33: Moving Autoroute 20 towards the north at the level of the Saint-Pierre Interchange (Excerpt from the proposal of dlandstudio, United States)

3.10

URBAN BOULEVARD

The conversion into an urban boulevard of a certain stretch of Autoroute 20 aims to facilitate the insertion of public transportation equipment, as well as active transportation into the highway right-of-way. Thus, images that illustrate urban boulevards always place forward a tramway or a cycling lane to illustrate the emphasis shift of the road equipment toward multifunctional equipment.

The automobile speed reduction on the sections retrofitted as an urban boulevard is perceived as a factor that will facilitate the inclusion of an urban highway. It also allows for the recomposition of the road experience through the planning of fast sections and then slower sections.

Finally, the urban boulevard also seeks to transform the identity perception of the highway equipment by planting an alignment of trees within the median lane. The trees thus give rhythm to a renewed landscape experience.

MAIN IDEAS
- Development of certain stretch of Autoroutes 20 and 720 as an urban boulevard

CRITERIA FOR DESIGN
- Highlight the sustainable development image of Montreal
- Ensure visual and landscape consistency of the highway features with the plant biomass of the adjacent territory
- Multiply transportation opportunities in the highway right-of-way
- Encourage planning and development of the neighborhoods adjacent to the highway
- Maximize the plant biomass to promote biodiversity
- Improve connections between neighborhoods by reducing barriers
- Minimize nuisances from the highway by reducing the speed of vehicles

Figure 4.34: Conversion of Autoroute 20 into an urban boulevard in the Turcot area (Excerpt from the proposal for Catalyse urbaine, Canada)

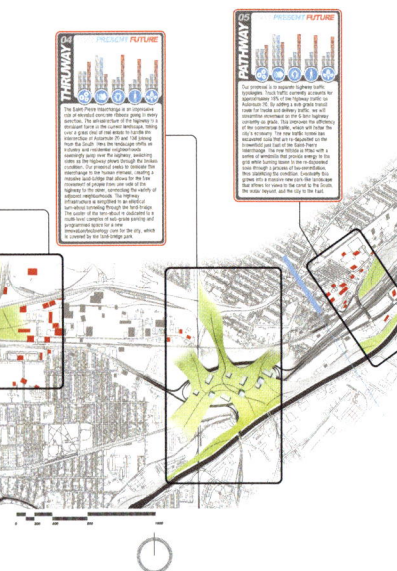

3.11

INTERCHANGES CONVERSION

The interventions that suggest the interchanges conversion into roundabout are based mainly on two aspects: the potential improvement in the traffic fluidity and the decrease of the highway right-of-way. The final objective of these two perspectives is to improve the quality of life for the neighborhoods adjacent to the highway by improving the accessibility and widening the buffer space between the highway and the living environments.

MAIN IDEAS
- The conversion of the interchanges into roundabouts is considered to rethink the relationship between regional and local traffic

CRITERIA FOR DESIGN
- Develop concepts of urban planning that improve traffic fluidity
- Increase development opportunities of the territory
- Redefine the relationship between local and regional use of the highway equipment

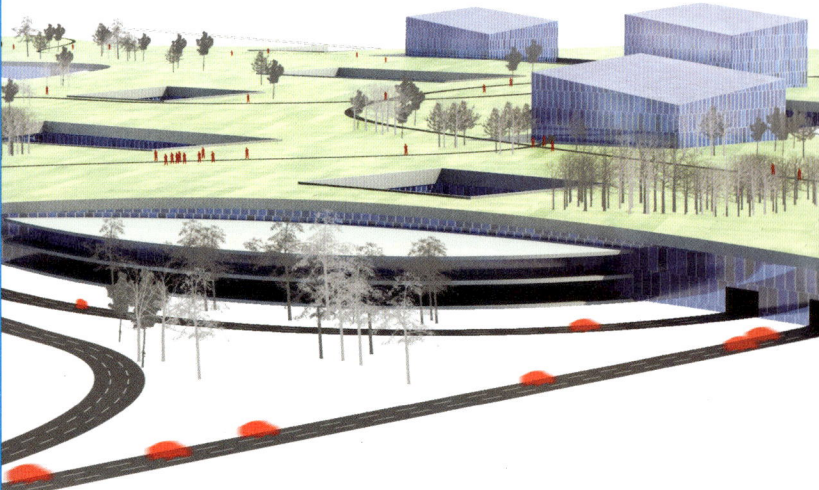

Figure 4.35 and 4.36: Conversion of the Saint-Pierre Interchange into an inhabited and partially covered roundabout (Studio Red, United States)

3.12

IMPLEMENTATION OF PUBLIC BUILDINGS IN THE HIGHWAY RIGHT-OF-WAY

Implementing new uses and new buildings to the residual spaces of the highway right-of-way is a strategy of visually marking the entrance pathways.

The buildings also play an animation role for the urban space with pedestrian and cycling routes that cross the highway space. The presence of buildings promotes the social animation of spaces that are usually empty and thus increases safety.

Beyond the landscape value of these interventions, it is nevertheless a strong concern for the multifunctionality of the territory that encourages these proposals. The use of residual spaces that is included in the highway right-of-way maximizes the use of the ground and promotes the city's sustainability as a whole.

MAIN IDEAS
- Implementing commercial buildings (hotels, shops) or parking decks under the interchanges' elevated structures and in the residual spaces of the highway right-of-way

CRITERIA FOR DESIGN
- Mark the pathway with strong and structuring visual landmarks
- Ensure the urban animation of pedestrian routes to increase safety
- Promote the territory multifunctionality to incorporate additional uses in the residual spaces of the highway right-of-way

Figure 4.37: Building that overlaps the lanes of the highway (Excerpt from the proposal of Brown and Storey Architects Inc.)

Figure 4.38: Infrastructure animation program (Excerpt from the proposal of Ghazal Jafari + Ali Fard, Canada)

CATEGORY 4
The territory, an element for road experience enhancement

4.1
DEVELOPMENT OF A LINEAR PARK/GREEN CORRIDOR

The recurring proposal of building a large natural park in the city entrance corridor is based on a radical redefinition of the landscape. From a gray and concrete image, the disaffected industrial areas become lush, animated places through wildlife and seasonal changes.

The presence of plants is also anchored on key elements of the landscape, like the Saint-Jacques Escarpment and the areas of residual vegetation located in the interstices between the different transportation infrastructures.

In addition to reinventing the highway and railway landscape, the creation of a linear park shows a strong potential for rethinking the interface of these infrastructures with the adjacent neighborhoods. It also allows to put in place new pedestrians and cycling routes as well as new experiences that take advantage of the winter, such as a cross-country skiing trails.

The designers also promote the ecological value of this intervention, which gives the possibility to connect several natural habitats on the island of Montreal This strategy is therefore built on a metropolitan planning perspective for natural corridors.

MAIN IDEAS
- Development of a linear park on the city entrance corridor

CRITERIA FOR DESIGN
- Promote the perception of seasonal changes through vegetation
- Highlight the sustainable development image of Montreal
- Create a network of green spaces that includes the Saint-Jacques Escarpment and the Lachine Canal
- Design a planted interface between the infrastructure and the living environments.
- Create new pedestrian and cycling routes
- Filter the visual nuisances of the highway and improve the air quality through vegetation

Figure 4.39: Conversion of industrial fields into a linear park (Excerpt from the proposal made by Moba Studio, Czech Republic)

Figure 4.40: Ecological corridor (Excerpt from the proposal of dlandstudio, United States)

4.2

DEVELOPMENT OF URBAN AGRICULTURAL AREAS AND COMMUNITY GARDENS

Disaffecting large industrial areas allows for the introduction of new uses on the territory. Several international experiences, for example in Detroit, used urban agriculture to suggest a new vocation to the territory. This transformation corresponds to both the multifunctionality and sustainability objectives through the promotion of local supply. These proposals are exploring various opportunities to exploit the Montreal territory for agriculture: the disaffected land, the roofs, and the community gardens. These different options are increasing the vitality of the areas that are crossed by the city entrance pathway.

Recalling the shape of the old agricultural plots allows for this new use to be anchored in a local historical reality, while promoting the identity of Montreal as a sustainable city through the implantation of urban planning ideas.

Designers also consider soil contamination to be a threat to farms and plan phytoremediation systems to gradually restore the soil quality to an acceptable level.

MAIN IDEAS
- Using brownfield and residual spaces for urban agriculture
- Establishing greenhouses dedicated to agriculture
- Developing community gardens

CRITERIA FOR DESIGN
- Highlight the sustainable development image of Montreal
- Preserve traces of the former agricultural plots divisions
- Diversify ground use by introducing agriculture as a potential urban use
- Decontaminate the industrial soil through phytoremediation techniques

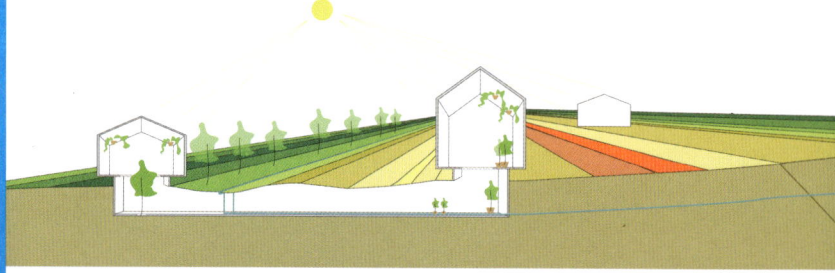

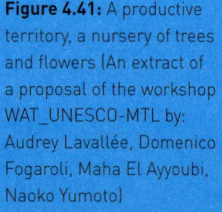

Figure 4.41: A productive territory, a nursery of trees and flowers (An extract of a proposal of the workshop WAT_UNESCO-MTL by: Audrey Lavallée, Domenico Fogaroli, Maha El Ayyoubi, Naoko Yumoto)

4.3

ENHANCING THE INDUSTRIAL HERITAGE

Chimneys, tall structures, and stacks of containers contribute to built landscape expressions and visual landmarks that testify to the historical and current industrial activities of this part of Montreal. Their use and update are designed to enhance not only the industrial identity of Montreal, but also the creative identity that is brought by industrial innovation.

MAIN IDEAS
- Creating art installations using containers as an element of expression
- Enhancing obsolete industrial structures that border the Lachine Canal

CRITERIA FOR DESIGN
- Highlight the creative image of Montreal
- Give a positive image to the industrial structures
- Create visual landmarks from the industrial structures of the territory
- Marking the territory with emblematic structures
- Designing a lighting scheme

Figure 4.42: Enhancing the emblematic bridge of Lachine (Excerpt from the proposal of David Garcia Studio, Denmark)

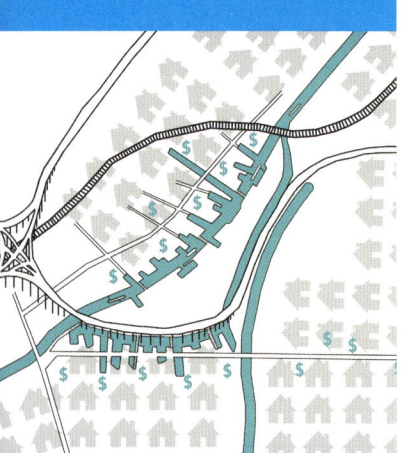

4.4
LACHINE CANAL ENLARGEMENT INTERVENTION

The pathway between the airport and downtown is the only entrance in Montreal that does not cross a bridge, thus making it difficult to perceive the island nature of the city. The presence of the Lachine Canal near Autoroute 20 is one of the only reminders of the importance of water and maritime transportation in the development of Montreal, although the canal is hardly visible. The proposals to enlarge the Lachine Canal aim to make the presence of water explicit on the entrance pathway to highlight the island character of Montreal.

In addition to affecting the quality of the landscape seen from alternative routes, the proposals also take into account the role of the canal as a pathway through the promotion of recreational aquatic activities.

The proposals also use water as a positive landmark in order to promote the redevelopment of disaffected areas. Secondary canals create an improved quality of life to attract a new population and revitalize neighborhoods.

MAIN IDEAS
- Enlarging some portions of the Lachine Canal
- Planning secondary canals

CRITERIA FOR DESIGN
- Recall the identity of Montreal as an island by giving a strong landscape presence to water
- Enhance the Lachine Canal
- Integrate the presence of water in all city entrance pathways
- Promote the revitalization and development of neighborhoods by using water as a positive landmark
- Diversify the natural habitats

Figure 4.43: The Lachine Canal extensions transform the Cabot area into an island. (Excerpt from a proposal of the WAT_UNESCO-MTL workshop by: Audrey Lavallée, Domenico Fogaroli, Maha El Ayyoubi, Naoko Yumoto)

Figure 4.44: View on the Lachine Canal extension from the highway (Excerpt from the proposal of Annexe U, Canada).

4.5

IMPLEMENTATION OF ENERGY PRODUCTION EQUIPMENT ON THE TERRITORY ADJACENT TO THE HIGHWAY

In order to highlight the sustainable image of the entrance pathway landscapes, these proposals explore the implementation potential, in the industrial areas and the interstices, of energy production infrastructures that use renewable energy sources, mainly the sun and the wind.

Located in under-valued areas (brownfield sites) or without vocation (roofs), energy production equipments are used to increase the vitality of the territory by using the energy industry that is already developed in other parts of the island with alternative sources of energy (e.g.: coal and gas in downtown with New City Gas and oil in the East of the island).

MAIN IDEAS
- Implementing wind turbines in the residual space of the city entrance corridor
- Installing solar panels on the edge of the highway or on the roofs of industrial buildings
- Using other energy sources to increase the vitality of the territory

CRITERIA FOR DESIGN
- Highlight the sustainable development image of Montreal
- Develop new energy production technologies adapted to urban environments
- Encourage the use of residual and undervalued spaces

Figure 4.45: Windmill park (Excerpt from the proposal of Gilles Hanicot, Canada)

4.6

OTTER LAKE RESTORATION

The Lachine Canal was dug from the enlargement and connection of an existing watercourse, including the St. Pierre River and Otter Lake, which was located at the foot of the Saint-Jacques Escarpment. This strategy aims to restore Otter Lake to enhance the scenic qualities of this place that mark the journey toward downtown.

The proposals reconsider the role of water in the city. From an ecological viewpoint, the proposals seek to steer run-off water toward Otter Lake in order to create a retention and filtration basin. From a functional viewpoint, the lake becomes an important recreational place.

MAIN IDEAS
- Developing a water basin in the center of the Turcot yard or on another site of the city entrance corridor
- Restoring wetlands
- Using this basin/wetland to collect and filter runoff water

CRITERIA FOR DESIGN
- Recall the identity of Montreal as an island by giving a strong landscape presence to water
- Give back to the former watercourses (Otter lake, Saint-Pierre River) a central urban presence
- Integrate the presence of water in all city entrance pathways
- Give access to new recreational activities (e.g.: water sports)
- Promote the revitalization and development of neighborhoods by using water as a positive landmark
- Diversify the natural habitats of the territory
- Use the bodies of water as buffer space between the infrastructure and the living environments

Figure 4.46: Lake in the Turcot rail yard (Excerpt from the proposal of Luis Callejas, Colombia)

Figure 4.47: New landscape seen from the highway (Excerpt from the proposal of DCYSA, Canada)

4.7

LIGHTING OR COLORING THE BUILDINGS ADJACENT TO THE HIGHWAY

Several buildings play an important role as visual landmarks in the city entrance landscape without having strong aesthetic or architectural qualities. The use of color allows for the enhancement of certain facades. Similarly, the lighting would create new impressions of the buildings. Together, these actions emphasize the creative image of Montreal.

The lighting also marks precise locations in the corridor, a few industrial buildings of interest or the downtown skyline, in order to enhance the perception of the landmarks. Thus, by acting on buildings, the lighting is used to script the progression toward downtown.

It should also be noted that lighting animates long winter nights. The use of winter lighting installations gives a perceptual quality similar to what is done in downtown during specific events (e.g.: Montréal en Lumière Festival).

MAIN IDEAS
- Coloring the facades of the industrial buildings along the highway
- Lighting buildings adjacent to the highway

CRITERIA FOR DESIGN
- Highlight the creative image of Montreal
- Use existing locations (industrial buildings, downtown skyline) to mark the city entrance pathways
- Create daytime and nighttime ambiances by developing concepts of innovative lighting and using new technologies
- Mark the pathway progression by creating visual landmarks
- Ensure consistency of night ambiances (e.g.: lighting scheme and coloring scheme).

Figure 4.48: Lighting industrial buildings (Excerpt from the proposal of TATINVESTGRAZGDAN PROJECT, Russia)

4.8

GREEN ROOFS IMPLEMENTATION

On several sections of the city entrance pathway, roofs play an important role in defining the landscape qualities. The targeted interventions on the roofs therefore have a strong potential to transform the landscape. Thus, by targeting the installation of green roofs, the designers aim to put forward a sustainable image for Montreal.

It must also connect this action to the new perception experiences of the landscape through Google Maps/Earth. With these tools, the aerial image of a city affects the identity perception for the observers.

Green roofs also have an ecological impact particularly by reducing heat islands. Beyond the image of the city, the designers also aim at a better integration between nature and architecture.

The potential uses of green roofs are also explored by the construction of an aerial boardwalk on the roofs of new buildings or their use as a potential site for agriculture.

MAIN IDEAS
- Installing green roofs on the buildings of the city entrance corridor

CRITERIA FOR DESIGN
- Highlight the sustainable development image of Montreal
- Promote the perception of a positive image of Montreal using new technology platforms (Google Earth)
- Encourage the use of residual and undervalued spaces
- Reduce the environmental impact of buildings (heat island, treatment of runoff water) and increase biodiversity

Figure 4.49: Green platform going over the highway and depositing on building roofs (An extract of a proposal of the workshop WAT_UNESCO-MTL by: Lin Juan, Selim Ayari, Emidio Arcidiacono, Valéry Simard)

Figure 4.50: Green roofs visible from Highway 720 (Extract from the proposal of TATINVESTGRAZG-DANPROJECT, Russia)

4.9

TOPOGRAPHY MODULATION/TRANSFORMATION

The Montreal landscape is marked by a large topographic element, Mount Royal. The Saint-Jacques Escarpment is one of the gradients that mark the ascent to the mountain. The interventions on the topography are intended to play with the original topographic components of Montreal.

In addition to creating a new landmark in the Montreal landscape, the presence of a new hill creates new viewpoints. In this sense, the new hills are new hiking or climbing routes depending on the slopes.

The development of hills also has the objective to find a solution to the production of excavated material during the construction of new infrastructure. Thus, the material produced as waste during works becomes positive markers for the landscape.

In an *ad hoc* manner, the hills can be used to create new interfaces with the neighborhoods adjacent to the highway in playing the role of noise-reducing wall.

Figure 4.51: New mountain near the Saint-Pierre district (Excerpt from the proposal made by Gerwin de Vries + Alexander Herrebout, The Netherlands)

Figure 4.52: Inhabited topography in the Saint-Jacques Escarpment (Excerpt from the proposal of Clément Boitel, France)

MAIN IDEAS
- Creating hills/mountains in the city entrance corridor
- Changing the topography of the city entrance corridor to reveal the singular elements or create ambiances

CRITERIA FOR DESIGN
- Create new perspectives on the Montreal territory
- Develop creative uses for the excavated material that is produced from the construction of infrastructure
- Use topography as a significant element for the territory

4.10

SNOW DUMPS IMPROVEMENTS

Snow is an essential component of Montreal's winter landscapes. These proposals are designed to increase the presence of snow in the landscape of the city entrance corridor by implementing snow dumps that create short-lived artificial hills.

The presence of snow would also develop some winter recreation activities.

Planning snow dumps also aims to develop an ecological role for the infrastructure by providing snow filtration and depollution during melting in the spring.

MAIN IDEAS
- Using snow dumps to create seasonal ambiances

CRITERIA FOR DESIGN
- Develop new expressions for winter landscapes
- Dot the entrance pathways with temporary winter features
- Promote winter appropriation for the city entrance corridor
- Reduce the environmental impact of snow dumps

Figure 4.53: Accumulation of snow (Excerpt from the proposal of Luis Callejas, Colombia)

Figure 4.54: Snow dumps and treatment of used snow (Excerpt from the proposal of LATERAL OFFICE, Canada)

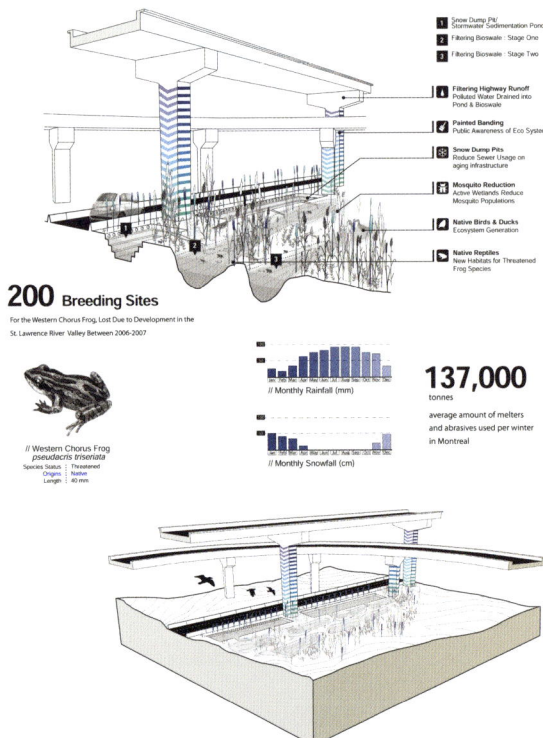

4.11

MOUNT ROYAL ENHANCEMENT

Mount Royal, as an emblematic element of Montreal, is relatively unnoticeable from the airport to downtown entrance corridor. These proposals are designed to protect the visual perspectives that allow a better recognition of the Mount Royal shape, especially from the airport field, as well as the creation of new opportunities by building towers or developing panoramic viewpoints.

This strategy would improve some mechanisms that protect viewpoints towards Mount Royal that currently exists (e.g.: City of Montreal Master plan).

MAIN IDEAS
- Protecting views toward Mount Royal from Montreal-Trudeau Airport
- Construction of a panoramic viewpoint overlooking Mount Royal

CRITERIA FOR DESIGN
- Preserve views toward Mount Royal
- Create scenic lookout to observe Mount Royal

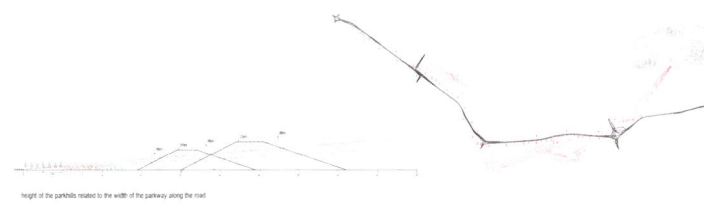

Figure 4.55: Relations between different topographic interventions and Mount Royal (Excerpt from the proposal made by Gerwin de Vries + Alexander Herrebout, The Netherlands)

4.12

URBAN RUNOFF WATER TREATMENT

A large number of watercourses, formerly present on the Island of Montreal, are today disappeared. These interventions seek to give a strong presence to water in the urban landscape of the city entrance corridor. Opportunities to do this are multiple and would create landscape wetlands, streams, and ditches.

These places have an important ecological role in collecting and filtering runoff water and creating new natural habitats.

The presence of water also would positively connote the urban environments to promote revitalization or development.

MAIN IDEAS
- Implementation of an open system to collect and treat runoff water

CRITERIA FOR DESIGN
- Highlight the sustainable development image of Montreal
- Give the former watercourses a strong urban presence
- Promote the revitalization and development of neighborhoods by using water as a positive landmark
- Use landscape as an infrastructure to manage runoff water

Figure 4.56: Using water in new developments (Excerpt from the proposal of Thibodeau Architecture + Design, Canada)

Figure 4.57: Management of runoff water from the Saint-Jacques Escarpment (Excerpt of a proposal of the workshop WAT_UNESCO Montreal by: Catherine Blain, Valeria Poggiani, Taiki Fujimaki, Andrea Spector)

Forest area
Parks
Water feature

4.13

RESIDENTIAL AND COMMERCIAL INTENSIFICATION

The residential and commercial intensification of the city entrance corridor is frequently suggested. In general, the promotion of a dense to very dense urban territory is sought.

Devitalized industrial areas are the main place for these new arrangements and the industrial heritage that is present form the basic landscape structure.

The urban planning strategies used to intensify refer to the creation of development poles and neighborhoods that have a distinctive individual identity. This concept also highlights the use of train stations as central node in the intensification strategies (e.g.: transit-oriented development).

The presence of high-rise buildings outside downtown also offers new viewpoints of the Montreal landscape since no elevated point of observation is currently present west of downtown.

The desired interface type for the different pathways also has an impact on the architecture of the buildings. On the edge of the Lachine Canal, the composition of the buildings promotes the animation of public space. On the edge of the highway and railway infrastructure, the visual impact is created through the volumetric composition.

MAIN IDEAS
- Creating residential and commercial (mixed) neighborhoods in the former Turcot rail yard or in the devitalized industrial areas
- Concentration of new developments in the vicinity of the Lachine Canal or public transportation stations

CRITERIA FOR DESIGN
- Promote the perception of an attractive and dense city
- Enhance the industrial heritage
- Animate the outskirts of the Lachine Canal
- Modulate the building's architecture as a function of the desired interfaces for the various pathways
- Intensify the territory arrangement while intensifying residential and commercial uses in the vicinity of public transportation stations and poles

Figure 4.58: Intensifying Norman Street industrial area (Excerpt from the proposal of Brown + Storey Architects Inc., Canada)

Figure 4.59: Intensifying the former Turcot rail yard (Excerpt from the proposal of dlandstudio, United States)

4.14
DESIGNING ICONIC BUILDINGS

These proposals seek to use architecture to create a new strong and distinctive visual landmark in the city entrance corridor. This strategy, called the "Bilbao effect" because of the construction of the Guggenheim Museum by Frank Gehry in the city of Bilbao in Spain, is now common in cities that seek to fit in the global network of architectural and artistic tourism.

At a local level, designers also saw the large brownfield sites as a favorable place for the implementation of large equipment that can be difficult to insert in urban environments.

The suggested uses are varied and target as much public equipment (hospitals, universities) as industrial or commercial buildings.

MAIN IDEAS
- Using large vacant spaces in the industrial areas (Turcot rail yard) to implement major equipment whose architecture would have an iconic character
- Building towers and high-rise buildings along the highway and railway path

CRITERIA FOR DESIGN
- Highlight the creative image of Montreal by designing an emblematic architecture
- Mark the pathway with strong visual landmarks
- Set up major urban equipment in the corridor free spaces

Figure 4.60: Megastructure covering Autoroute 20 (Excerpt from the proposal of Brown & Storey architects, Canada)

Figure 4.61: Reference building over the highway (Excerpt from the proposal of ZerOgroup + Fabrica Paisaje, Brazil)

4.15

IMPLEMENTING COMMERCIAL AND PUBLIC SERVICES

These proposals seek to give anew function to brownfields in need of revitalization. Recycling old industrial buildings is at the heart of several intervention strategies to enhance their heritage value. Similarly, public and commercial uses allow these areas to remain important employment poles. These ensure the presence of attractive elements that convert the various areas of the pathway into a place of destination rather than simple transit spaces.

To achieve this revitalization, several strategies rely on the presence of transit stations to structure the development poles.

MAIN IDEAS
- Diversifying uses in the brownfields by inserting buildings with public purposes (sports center, schools, health services, public market) or commercial (hotel, business center, office space)

CRITERIA FOR DESIGN
- Reuse and recycle industrial buildings to accommodate new uses
- Develop employment centers and public services
- Intensify the territory next to public transportation stations

Figure 4.62: New university campus (Excerpt from the ANG proposal, United States)

Figure 4.63: Public equipment along the highway in the St. Pierre area (Excerpt from a proposal of the workshop WAT_UNESCO Montreal by: Christine Robitaille, Claudio Mura, Bian Xiaozhe, Ahmad Almahairy)

4.16

USING CULTURE AS A SOURCE OF REVITALIZATION

The use of culture and artists to start the process of urban revitalization is a common strategy for several cities. In the context of the city entrance corridor, it has three different meanings. On the one hand, if the routes offer a view of the Montreal landscape, the places that make it up can also become destinations to enjoy the local knowledge, such as museums or exhibition centers. Then, artists are usually ready to occupy marginal spaces more accessible on a financial level. The city entrance corridor has several spaces in the brownfield sites. The presence, and especially the action, of artists would lead to the production of new landscape expressions. The existing places for industrial production can also become production places for objects coming out of the design approach for designers.

MAIN IDEAS
- Implementing artistic or heritage distribution center (art gallery, design center, museum)
- Planning artistic or design workshops to encourage the emergence of a creative class in the industrial zones of the city entrance corridor

CRITERIA FOR DESIGN
- Highlight the creative image of Montreal
- Promote quality processes in design (design competition, workshop)
- Use culture and design as a revitalization vector for the brownfields
- Promote culture as a lever for sustainable development

Figure 4.64: Artist workshop in the St. Pierre area (Excerpt of a proposal of the workshop WAT_UNESCO Montreal by: Christine Robitaille, Claudio Mura, Bian Xiaozhe, Ahmad Almahairy)

Figure 4.65: New museum in the former Turcot rail yard (excerpt from the proposal of TATINVESTGRAZG-DANPROJECT, Russia)

4.17

REVITALIZATION OF THE INDUSTRIAL AREAS

The historic multiplicity of the transportation routes in the city entrance corridor strongly contributed to build the current industrial landscape. Several designers identified the revitalization of the industrial areas as a strong idea to promote a new territorial vitality. This strategy would promote reusing inherited industrial buildings and preserve this historical use of this territory.

The designers suggest updating the production activities that are there, notably through establishing industries in the ecological field, involved in recycling or operating renewable energy sources.

Renewing the industrial areas is also suggested by intensifying the territory density. While the multiple floor industrial buildings were common around 1900, this type of building is rather short today. Encouraging the presence of several level industrial buildings would make this land arrangement more noticeable in the city entrance pathways landscapes.

MAIN IDEAS
- Revitalizing industrial areas by implementing new production activities (energy, recycling)
- Rearranging and intensifying industrial areas

CRITERIA FOR DESIGN
- Enhance the industrial heritage through preserving buildings and uses
- Improve the positive perception of the industrial nature from the city entrance pathways
- Intensify the arrangement of the territory by increasing and diversifying the presence of the industrial area

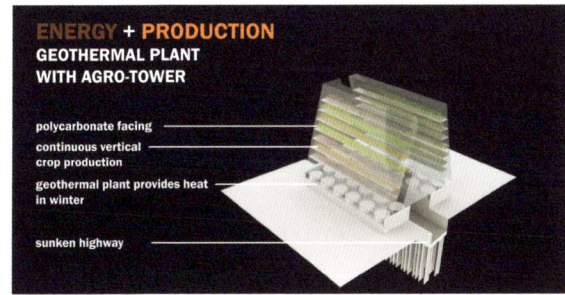

Figure 4.66: Revitalization of industrial areas through geothermal energy and other sources of clean energy, greening, and residential and commercial intensification (Excerpt from the proposal of dlandstudio, United States)

4.18

RE-PLANNING OF THE MONTREAL-TRUDEAU AIRPORT SITE

The Montreal-Trudeau airport is an enclave that generates a relatively limited urban presence in the Autoroute 20 gateway corridor. The orientation of the take-off/landing runways currently ensures that planes flying over are mainly visible from Autoroutes 520 and 40. Thus, arranging runways according to Autoroute 20 would give a greater urban presence in the Autoroute 20 corridor, nevertheless creating nuisance that should be reduced to preserve the quality of living environments.

In addition to the runways, the Montreal-Trudeau Airport can ensure a greater urban presence by redeveloping its outdoor spaces to promote accessibility to its terminal by methods of transportation different from the car. Thus, links with stations (bus and train) and additional pedestrian/cycling lanes would diversify the access and develop new potential uses.

MAIN IDEAS
- Relocating landing/takeoff runways
- Redeveloping receiving areas

CRITERIA FOR DESIGN
- Improve the urban presence of the airport

Figure 4.67: Boarding platform of the airport (Excerpt from the proposal of DCYSA, Canada)

Figure 4.68: Alignment of landing runways and highway (Excerpt from the proposal of Clément Boitel, France)

CATEGORY 5
The territory intrinsic strategies

5.1
CREATION OF PARKS

Creating a park or a new public space is often a key moment in revitalization projects since it renews both the quality of life and the image of the neighborhoods in which these actions are taken. First, it is in this perspective that the creation of new parks is considered.

The second strategy after park creation is developing welcoming spaces near railway stations. On a large scale, this welcoming public space is also suggested for the terminal. A place that would allow for the observation of planes taking off.

MAIN IDEAS
- Creation of neighborhood/local parks

CRITERIA FOR DESIGN
- Promote quality processes in design (design competition, workshop)
- Create significant entrance and exit points for all entrance pathways
- Stimulate the revitalization of neighborhoods through the construction of friendly public spaces and parks
- Focus on an inclusive approach for sustainable development

Figure 4.69: Development of neighborhood parks near a station (Excerpt from the proposal of Gilles Hanicot, Canada)

Figure 4.70: Punctual park at the Lucien-L'Allier station (Excerpt of a proposal of the workshop WAT_UNESCO-MTL by: Mazen Kandalaft, Marie-Pierre McDonald, Sara Saadouni, Khuplianlam Tungnung)

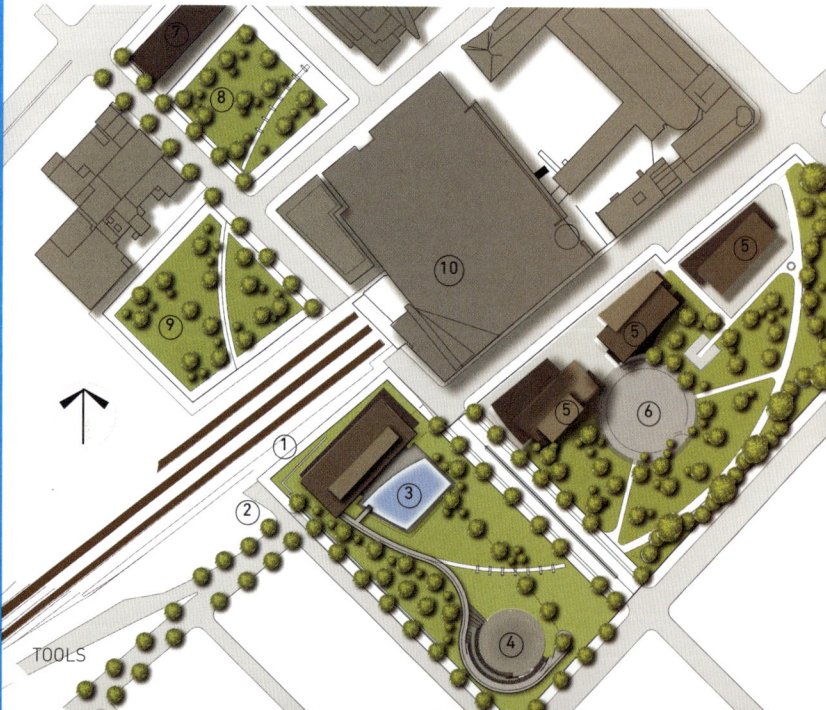

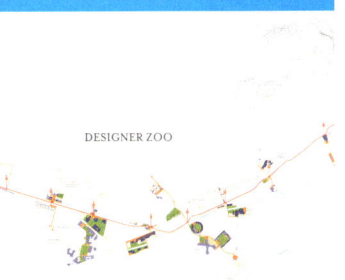

DESIGNER ZOO

5.2

IMPROVEMENT/ADDITION OF PUBLIC TRANSPORTATION LINES

Although the highway plays an essential role for the city entrance corridor, several proposals sought to give public transportation a comparable urban presence. All transportation types (train, tram, bus, trolleybus) were considered.

The connection between the airport and downtown was a major concern for the designers. If suggesting an effective pathway was a goal, the proposals also aimed to offer viewpoints towards the Montreal neighborhoods. Thus, the suggested airport/downtown connections used as much of the existing the rail tracks as they did new routes through neighborhoods, including the revival of old railway tracks that would highlight the island character of Montreal.

Improving the public transportation services is also considered as a strategy to revitalize neighborhoods.

MAIN IDEAS
- Commissioning a rail service to link airport/downtown
- Implementing tramway lines for local services
- Implementing electric bus lines
- Developing a public transportation system by hot-air balloon

CRITERIA FOR DESIGN
- Diversify and improve the public transportation offerings particularly linking the airport/downtown
- Stimulate the neighborhoods revitalization using public transportation as a lever for development

Figure 4.71: Improving the airport/downtown connection (Excerpt from the proposal made by Arnaud EFOE Architect, France)

Figure 4.72: Adding public transportation lines (Excerpt from the proposal of David Garcia Studio, by Denmark)

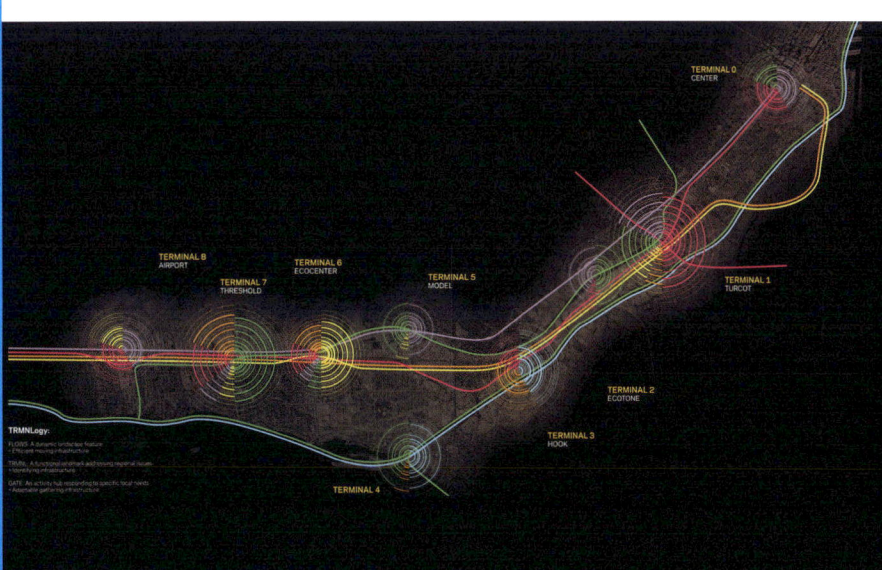

5.3

RENOVATION AND/OR ADDITION OF LOCAL STATIONS

The importance given to public transportation is accompanied by an attention given to the design and location of transit stations. These stations are perceived as the ends of the individual pathways through the city entrance corridor, explaining the importance of creating meaningful places.

The presence of stations also reflects the possibilities of exchanges between the methods of transportation. In this sense, stations are also nodes in individual pathways. One of these urban nodes, the Saint-Pierre interchange, is targeted as a privileged place to implement a new station.

The presence of stations also seeks the development of an urban development strategy focused on an intensification of the activity in their vicinity.

MAIN IDEAS
- Implementing new railway stations
- Improving the quality of waiting areas and public spaces located around the railway stations
- Creating arrangements that promote intermodal transit

CRITERIA FOR DESIGN
- Build stations with great architectural and urban quality to reflect their role as a neighborhood entrance
- Use stations as a place of intermodal exchange
- Stimulate the neighborhoods revitalization using public transportation as a lever for development

Figure 4.73: New line of train, train station, and new public space (Excerpt from a proposal of the workshop WAT_UNESCO-MTL by: Christine Robitaille, Claudio Mura, Bian Xiaozhe, Ahmad Almahairy)

Figure 4.74: New intermodal station in Dorval (Excerpt from the proposal of Brown + Storey Architects Inc., Canada)

5.4

MODIFICATION OF THE LOCAL STREETS NETWORK

Crossing the city entrance corridor by car is usually only considered from the highway. Designers identified the local streets network as an essential component of the corridor since it offers other experiences usually closer to the crossed neighborhoods.

The predominant objective in developing new streets or modifying existing ones is an increased urban permeability. One of the chosen strategies aims at diversifying route options by linking the airport with downtown. In addition to multiplying landscape experiences in this area, this option also aims to offer alternative routes to improve the fluidity in case of road work or traffic jams.

The designers also target a greater permeability at the local level. Thus, the suggested axes would connect the neighborhoods and cross the infrastructures.

MAIN IDEAS
- Planning new streets
- Extending or modifying the alignment of existing streets

CRITERIA FOR DESIGN
- Improve the territory permeability by diversifying the potential vehicular routes
- Improve the permeability between neighborhoods by developing new vehicular crossing for the infrastructures
- Develop an integrated approach of the shared street including the environment, the mobility, and the local culture

Figure 4.75: Realignment of the boulevard (Excerpt from the proposal of Gilles Hanicot, Canada)

Figure 4.76: Transformation of the Cabot area in an island with limited local streets (Excerpt of a proposal of the workshop WAT_UNESCO-MTL by: Audrey Lavallée, Domenico Fogaroli, Maha El Ayyoubi, Naoko Yumoto)

5.5

COMMISSIONING TAXI-BOATS

Using the Lachine Canal as a city entrance pathway remains only an anecdotal and recreational event today. The designers wish to give the Lachine Canal and the St. Lawrence River a major role in linking the airport to downtown by commissioning water taxis or cruise shuttles. Creating a canal connecting the river to the airport along Dorval Avenue would connect the airport more effectively.

This new experience of city entrance puts the island character of Montreal forward and increase visibility of the entire industrial complex of the Lachine Canal.

MAIN IDEAS
- Establishing a ferry/taxi-boat service along the Lachine Canal
- Creation of new canals and wharves to allow boat access

CRITERIA FOR DESIGN
- Recall the identity of Montreal as an island by giving a strong presence to water
- Build new welcoming public spaces at the airport to promote the identity of Montreal
- Highlight the role of the Lachine Canal as a place for maritime transportation
- Create new pathways using water bodies

Figure 4.77: New canal along Dorval Avenue (Excerpt from the proposal of Tokyo Institute of Technology Tsukamoto Yoshiharu lab., Japan)

Figure 4.78: Taxi-boats and kayaks on the Lachine Canal (excerpt from a proposal of the workshop WAT_UNESCO-MTL by: Valérie Gravel, Caroline Cajelais, Rève Aoun, Kohei Kobayashi)

5.6

CREATION OF FOUNTAINS

Water in an urban environment is a positive landmark of the public space. Designers used this characteristic in order to promote the creation of quality urban spaces.

In addition to a visual feature, the designers have also worked on the potential of water to create ambiances by developing fountain or "misters". In this perspective, water would create new interfaces with infrastructures by filtering the noise.

MAIN IDEAS
- Installation of public equipment that uses water (fountains, misters, jets)

CRITERIA FOR DESIGN
- Create new pathways using water bodies
- Filter nuisances from the transportation infrastructures (noise) by using water

Figure 4.79: Waterfalls along the Lachine Canal (Excerpt from the proposal of David Garcia Studio, Denmark)

Figure 4.80: Presence of water in urban areas (Excerpt from the proposal of DCYSA, Canada)

5.7

DEVELOPMENT OF MULTIFUNCTIONAL TRAILS (TERRITORY)

Montreal is already a cycling city equipped with a vast network of cycling lanes. Designers built on this identity by developing and consolidating the already existing network.

The suggested trails aim to promote the airport/downtown link by taking advantage of the highway or the railway linear axis. They are designed to facilitate the connection of neighborhoods, parks, green spaces, and other attractive locations (e.g.: Lachine Canal, shores of the St. Lawrence River, Saint-Jacques Escarpment).

The multifunctionality of these lanes is also developed to accompany the cycling lanes with pedestrian, cross-country skiing, or horse riding paths.

Designers also developed a uniform visual image for the whole network, an image defined by the signage and lighting system.

MAIN IDEAS
- Developing multifunctional trails (cycling lanes, pedestrian paths) in the whole territory of the city entrance corridor
- Consolidating the existing cycling lanes network
- Building elevated multifunctional lanes

CRITERIA FOR DESIGN
- Highlight the sustainable development image of Montreal
- Create daytime and nighttime ambiances by developing concepts of innovative lighting and using new technologies
- Develop new opportunities to use this network for other recreational activities
- Design a uniform signage system for cycling lanes on the territory of the entrance corridor

Figure 4.81: Bike lanes at night (Excerpt from the proposal of Johnathan Yue, Singapore)

Figure 4.82: Multifunctional lanes (Excerpt from the proposal of David Garcia Studio, Denmark)

5.8

UNDERGROUND PATHWAY

The idea of creating new types of pathways was considered by several designers. The underground pathway is one of these new experiences of city entrance. However, it is built on strong identity and historical elements of Montreal: the presence of buried watercourses which could continue to be rediscovered as well as the image of underground Montreal in downtown.

MAIN IDEAS
- Creating an underground pathways in the municipal aqueducts

CRITERIA FOR DESIGN
- Create new pathways

Planning Scenarios

Beyond identifying design criteria associated with each intervention strategy, the planning scenarios are intended to consolidate the strategies that respond to different planning visions of the city entrance territory. In fact, by creating a coherent representation of a possible future, the scenarios offer the prospective images that seek to make explicit the large choice of layouts that are shown to the stakeholders on the Autoroute 20 gateway corridor.

The definition, description, and illustration of the planning scenarios were carried out by grouping the intervention strategies that offered convergent planning principles and that were dependent on one another in a perspective of territorial coherence. The five planning scenarios address the following themes:

1. **A dense transit-oriented corridor**

2. **An ecological and recreational corridor**

3. **A culture and heritage corridor**

4. **A corridor along water**

5. **A productive corridor**

SCENARIO 1
A dense transit-oriented corridor

YUL/MTL

Figure 4.83: MNTRL TRMNL (David Garcia Studio, Danemark)

The Autoroute 20 gateway corridor between Montreal-Trudeau Airport and downtown is one of the largest potential areas for redevelopment of the whole island of Montreal. Its close proximity to downtown, as well as the presence of major public and recreational equipment, is a major asset that promote this redevelopment. Ensuring the vitality of the city entrance corridor is reflected first by a dynamic occupation of the territory in order to make a lively and friendly living environment.

A dense and compact city
The redevelopment models suggested by the designers are mainly based on the creation of a compact, dense, and mixed city. However, the conditions to implement this type of development are based on several transformation factors of the territory. The first factor is transforming the industrial zones into living environments, a transformation that often requires territorial planning and soil decontamination processes. The areas covered by this transformation are mainly the former Turcot rail yard as well as other devitalized industrial areas, such as Norman Street, the LaSalle industrial area in the south of the Lachine Canal, as well as the Dorval and Lachine industrial areas, north of Autoroute 20.

Improvement of public transportation
An essential tool in promoting a dense urban corridor is the improvement of the public transportation services, especially the addition of high capacity lines (train, tram), and the renovation or addition of railway stations. This transportation equipment creates new polarities in the territory that increase the potential for the development of a mixed urban environment. Unlike the intensification process itself that affected the transformation of landscape expressions perceived from the highway, strategies to renew public transportation are mainly intended for neighborhoods and only indirectly consider the highway experience.

The creation of an airport/downtown link tops the designers' list of concerns related to public transportation, most of which proposed to use the CP or CN current line. Stations are mainly added in the industrial area of the Norman Street and near the former Turcot rail yard. These two new stations are used as a development catalyst to redefine the adjacent environments. The Dorval, Vendome and Lucien-L'Allier stations are also renovated to allow a better intermodal link, ensuring their welcoming role.

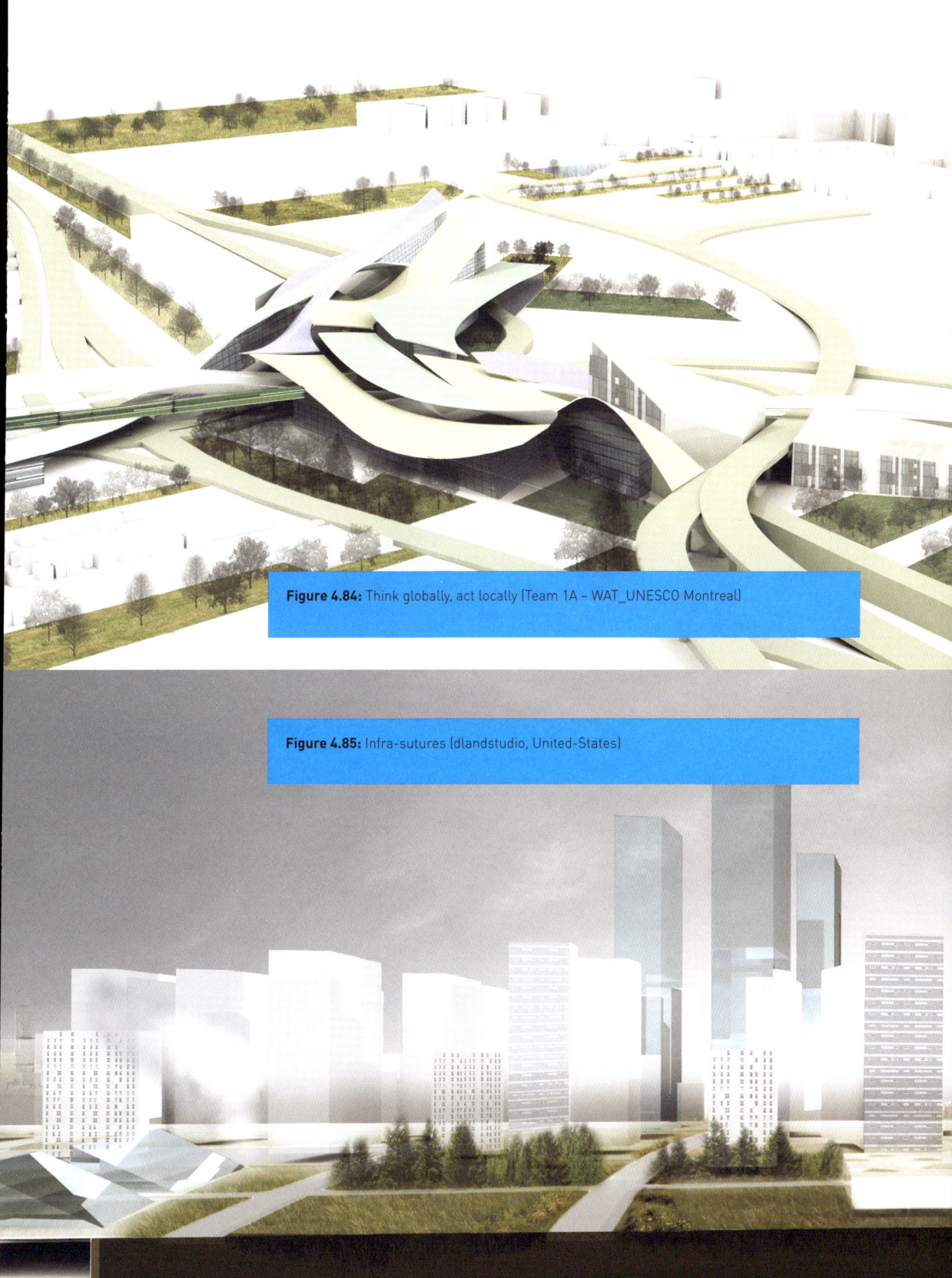

Figure 4.84: Think globally, act locally (Team 1A – WAT_UNESCO Montreal)

Figure 4.85: Infra-sutures (dlandstudio, United-States)

Prospective scenarios make coherent representations of a possible future for the territory.

Amongst other transport lines that are implemented in the proposals, one is located in the Bouchard Blvd in Dorval, the Victoria Street in Lachine, and extends to the Notre-Dame Street West in the Sud-Ouest Borough. This route adds a local perspective to the city entrance pathway since it passes through the historic corridor for developing the west of Montreal. In addition to contributing to the revitalization of Lachine, this line contributed to redeveloping the industrial area located between 1st avenue and the Saint-Pierre interchange. This area also welcomes in a few proposals concerning the idea of a new station on the Candiac line commuters' train (AMT).

Finally, the St. Patrick Street is also granted with a new public transportation infrastructure. This line promotes the redeployment of industrial areas south of the Lachine Canal.

Increase in permeability

The third essential element in the promoting the dense urban corridor is increasing the permeability on the whole territory. Proposals that concern this transformation perspective are based on an important finding: the presence and multiplicity of transportation infrastructure (road and rail) create barriers that limit crossings and therefore increase the route time. Interventions to increase the permeability are designed to increase accessibiliy to the public transportation nodes and indirectly promote the density.

A lighter response to the problem of barriers, given by the designers, is the construction of pedestrian passageways. They are mainly located at the intersection of Autoroute 20 and Dorval Avenue, off of 32nd avenue in Lachine, in the Saint-Pierre district, in the former Turcot rail yard, in the Turcot Interchange, as well as at the level of the Ville-Marie Expressway. Each time, the gateways are used to connect the existing neighborhoods through railway stations or meaningful places of the territory such as the Saint-Jacques Escarpment or the Lachine Canal. When they are wider, the gateways become elevated parks, particularly close to the Saint-Pierre district, as well as along the Ville-Marie Expressway to create a belvedere.

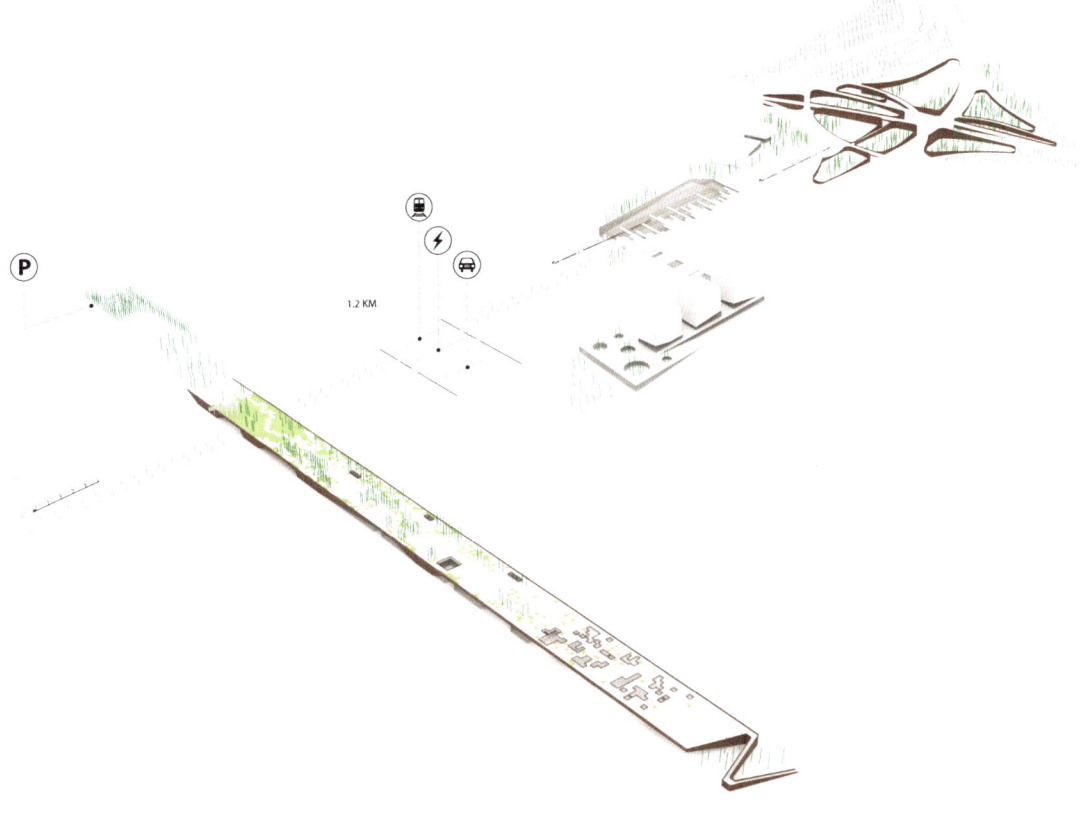

Figure 4.86: Passive >> Performative (Ghazal Jafari + Ali Fard, Canada)

Heavier interventions are suggested by designers to reduce barriers. They are more directly aimed at the highway by moving its footprint, its burying, or the construction of a megastructure that incorporates multiple methods of transportation as well as pedestrian crossings. These interventions are used to eliminate or combine barriers to reduce their overall impact. They are mainly suggested for West-Montreal as well as in the former Turcot rail yard.

Finally, the third type of response concerning the improvement of the permeability is linked to the quality of animation in the public space. To promote the movement of pedestrians, the access canals to the stations must have a minimum animation especially when crossing unfriendly places such as the industrial areas and the highway. To do so, several public occupation strategies for the highway are suggested where structures are elevated, more particularly under the Saint-Pierre and Turcot interchanges.

SCENARIO 2
An ecological and recreational corridor

YUL/MTL   TOOLS

Figure 4.87: Ecological and recreative right-of-way (Tokyo Institute of Technology Tsukamoto Yoshiharu Lab, Japan)

The presence of vegetation in the city entrance corridor as perceived from the highway is relatively low and located in the background. The foreground is dominated by constructed, paved, or concrete surfaces, which show a low quality ecological environment. The scenario that concerns the creation of an ecological and recreational corridor aims to reverse this image and give nature a leading role in the whole city entrance.

Greening the highway
The first component of this action is greening the highway itself. In addition to renewing the landscape perception from the highway route, this action allows for enjoyment of the benefits of vegetation, in particular that of filtration and a redefinition of the interface with the adjacent living environments. For the highway right-of-way, designers suggest inserting vegetation in the freed-up interstices of the highway as well as under the elevated structures. Planting trees is suggested to redefine the visual frame for the road users, as well as to give a seasonal perspective through the changing foliage. The highway structures are themselves subject to vegetation through the development of a vertical planting strategy on the noise-reducing walls as well as the pillars of the elevated sections.

To amplify the vegetation-filtering role, some designers also suggested creating mounts along the highway. More or less elevated, these mounts give new landscape perspectives by making vegetation more important in certain places, especially in the rectilinear hallways of Autoroute 20 like in Dorval and Lachine or in the former Turcot rail yard.

Ecological corridor
A second component of this scenario is the creation of ecological corridors to connect the existing green spaces to the city entrance. These corridors were mainly intended to allow the fauna and flora to disperse in the territory thus increasing the biodiversity of the environments.

This action also allows better development and maintainance of the rich ecological environments that are currently present through the creation of an adjacent environment that is favorable to their development.

The Saint-Jacque Escarpment, Lachine Canal, areas of residual vegetation, and the shores of the St. Lawrence River thus are the anchor points for ecological corridors. In these spaces, existing large parks or green spaces are included such as the Angrignon Park and the Meadowbrook Golf Course. To connect these places, large portions of industrial areas are re-naturalized and portions of Autoroute 20 are buried.

These great greening initiatives lead to a radical transformation of the city entrance corridor. From all pathways, landscape perceptions put forth natural spaces. On a smaller scale, other actions accentuate these perceptions such as greening rooftops in the industrial area of Dorval and Lachine as well as in the Saint-Henri district. New local parks also contribute to punctuate lively places along the corridors.

Multifunctional trails network

Finally, the presence of ecological corridors also allows the development of a vast network of cycling lanes and pedestrian trails. These trails are inserted within the highway right-of-way to provide fast lanes for bicycles heading toward the existing trails in the neighborhoods in order to serve the local population. In broader natural environments, trails are used as a place to explore and observe the fauna and flora of places like the Saint-Jacques Escarpment.

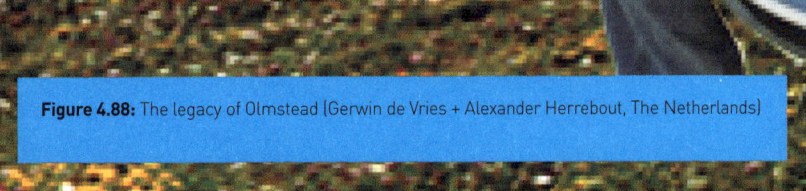

Figure 4.88: The legacy of Olmstead (Gerwin de Vries + Alexander Herrebout, The Netherlands)

SCENARIO 3
A culture and heritage corridor

YUL/MTL

Figure 4.89: MNTRL TRMNL (David Garcia Studio, Denmark)

The city entrance concept is parallel in many ways to a welcoming center, where the visitor is given a first impression of Montreal's local identities. These are expressed as much by the peculiarities of the landscapes as by the shared adherence to certain values that articulate and direct the local project. The proposals for improvements suggested for this scenario are based on this vision of the city entrance and seek to seize all opportunities to stage Montreal identities and, more particularly, its creative nature.

Artworks
The first action related to this scenario is to insert, in the city entrance pathways, the product of cultural nature such as artworks. These are implemented as autonomous elements on the pathway, creating identity markers. The location of the artworks is either distributed on the entire highway or dedicated to specific locations representative of tension points in the progression toward downtown, including the interchanges. In this respect, the Dorval and Turcot interchanges are privileged places to implement artworks.

The artworks also take different forms of expression than the autonomous piece. Often, they invest and integrate the infrastructures and buildings through light projections, frescoes or new cladding materials. In this context, the artistic interventions alter the perception of the built landscape as much from the highway as the other pathways. The main elements of the highway covered by this type of intervention are the lighting structures, noise-reducing walls, as well as the pillars and roadway decks. In the latter case, the structure's elevation allows for the possibility to create strong nocturnal visual landmarks from every pathway.

In neighborhoods adjacent to the highway, it is mainly the historical industrial structures, the blind factory walls located close to the highway, as well as the profile of downtown, that are targeted for light interventions.

Signage system
Cultural interventions also aim at emphasizing the local identity peculiarities. They then have the goal of developing a signage system to better read the local attractions of each neighborhood. These signage systems may be multiple. In a more direct way, a signage system is developed by means of display elements that indicate directions to follow. One strategy developed in this sense uses the analogy of the terminal to identify important places of the city entrance corridor and invent a path to link these places.

Other direct signage strategies aim to display logos or events for informational purposes in the way of banners and flags installed in downtown. Designers also suggest using light or projection boards on the highway structures. These display methods were suggested all along the highway route.

Figure 4.90: MNTRL TRMNL (David Garcia Studio, Denmark)

More indirectly, a signage system is also developed through establishing visual landmarks or protecting visual perspectives (e.g.: toward Mount Royal) that facilitate the legibility and intelligibility of the territory. Here, highway exits are mainly targeted for the implementation of built landmarks. Signage items indicate the presence of many neighborhoods, districts, and cities in the west of the island of Montreal.

Revitalization program through cultural industries
Finally, culture and creativity are also used as a vector for local development in neighborhoods, in this case it is used mainly in the industrial neighborhoods to be revitalized. Establishing artists/designers workshops or exhibition centers is a mean used to facilitate the production and dissemination of creation products.

In the former Turcot rail yard and near the Lachine Canal, the suggested use of culture as a source of revitalization is centered on the implementation of cultural venues for the promotion of local art and heritage. Conversely, in the major industrial areas of Lachine, the rather old industrial buildings are reinvested through activities of cultural creations. Therefore, the potential of large buildings is used ,along with easily adaptable structures, to insert new activities. Small studios/workshops are also being built on the rooftops of factories and other industrial buildings to develop a mixture of these areas by layering functions. These new high-rise constructions change the perception between the places of creation and the places of industrial production.

SCENARIO 4
A corridor along water

YUL/MTL

Figure 4.91: Paysage fluide (DCYSA, Canada)

Proposals based on using water to renew landscapes of the entrance corridor are first anchored in an ecological and environmental reading of urban planning issues, then in a historical perspective, since several watercourses which no longer existing today used to cross this territory, particularly the Otter Lake.

Runoff water management
Several proposals suggest developing a collection and filtration systems for runoff water to reduce the impact of storm water on sewage facilities, as well as to improve the quality of the water that is discharged into the environment. At the same time, these interventions would insert ecological environments in the territory of the city entrance corridor to increase local biodiversity.

The insertion of landscape ditches and retention ponds is considered as much for the highway infrastructure, especially the Dorval, Saint-Pierre, and Turcot interchanges, as for local streets in Dorval, Lachine, and Montreal West. Thus, a network of ditches gradually deploys over the highway and railway paths towards the living environments to finally reach existing bodies of water such as the Lachine Canal or the shores of the St. Lawrence River. Through this itinerary, the ditches make visible the ecological interactions and continuities of the territories they pass through.

To this more natural network, interventions are also added that are intended to provide greater symbolic presence to runoff water, particularly by taking advantage of the escarpment to create temporary falls visible from Autoroute 20.

Resurgence of buried watercourses
This new presence of water in the city would also recover local heritage through the resurgence of watercourses that are buried today. Although water is an essential component of Montreal identity, the lack of access to the shores as well as the burying of several watercourses erased the multiple historical links the city had with water. Increasing water presence in the landscape allows for a re-reading of this essential character.

As mentioned above, the Otter Lake located in the former Turcot rail yard as well as a small river (Saint-Pierre), which flowed from the Meadowbrook Golf Course toward the former Turcot rail yard, form the main disappeared watercourse that are being recreated by the designers.

Figure 4.92: The time-machine (Clément Boitel, France)

Revitalization through the use of water as a positive connotation factor on the landscape level

The revitalization of living environments is also an important urban planning issue that is considered through increasing the water presence in the city entrance corridor. Water allows to positively connote the local landscape and, thus, to promote the territory redevelopment. Widening the Lachine Canal is the main intervention suggested in this sense. Through the creation of secondary canals, new territories acquire a direct access to the water, which creates different, high-quality living environments. This strategy is used as much to enhance the existing environments as to generate new developments. In the latter case, the design of buildings is linked to the development of a lively waterfront to ensure a friendly public space.

Water is also used in a more timely fashion to create neighborhood ambiances. Adding fountains would create a background noise that filters more unpleasant noises that are generated by the transportation infrastructures or industrial environments. This strategy is used in downtown as well as along the Lachine Canal and close to the Turcot and Saint-Pierre interchanges.

Finally, the presence of water ponds opens new potentials for developing recreational activities that increase the quality of life. Therefore, the water bodies fit into the general use of parks and green spaces in the city entrance corridor.

SCENARIO 5
A productive corridor

YUL/MTL

Figure 4.93: YUL/MTL (Gilles Hanicot, Canada)

The development of the Autoroute 20 gateway corridor is historically marked by the joint presence and interaction of industries and transportation. Well before inserting the highway, the presence of the canal and railway tracks promoted the emergence of an industrial urban fabric, which, today, is officially recognized in Canada as a national historic site. The competition of road transportation over rail transportation, the construction of the St. Lawrence Seaway, as well as the decrease in the hydraulic energy production from the canal locks jointly led to the gradual decline of older industrial areas in the eastern portion of the corridor. This industrial revitalization led over the last few years to the rehabilitation of buildings for residential and commercial use in most urban zones accessible from public transport areas, but large portions of the industrial territory still await revitalization. Could the industrial activity reinvest the city entrance territory in a significant way?

Revitalization of the industrial activity
Several designers perceived the need to revive the industrial nature of the city entrance corridor as a source of landscape vitality. This source of renewal for the industrial identity is in the development and establishment of new production lines that promote sustainable development, including the implementation of recycling and reusing centers. This strategy would anchor the sustainability of the territory through industrial production.

Renewing industrial production creates visual effects on the city entrance pathways in two ways. The first is the construction and rehabilitation of buildings. The contemporary architectural expressions would have a positive impact on the landscapes perception. The second strategy is the enhancement of existing industrial structures, such as lighting the chimneys and cranes, or planning the containers organization. The promotion of a strong visual presence for the containers is often suggested to increase the perception of the corridor's industrial nature.

These strategies are mainly suggested for the industrial areas of Dorval and Lachine. In these areas, the shape of the industrial urban fabric is often reviewed to promote its intensification. In the Turcot rail yard, enhancement strategies for industrial activity are normally limited to heritage elements.

Figure 4.94: Welcome to Montréal (K2T2R, Singapore)

Energy production from renewable energy sources
The productive corridor concept does not only develop from the already existing industrial activities. Energy production from renewable sources (wind, sun and geothermal energy) is also considered. In addition to visually connecting the territory and the transportation infrastructure, energy production would establish functional interactions. Industrial environments and transportation infrastructures are important consumers of energy. Energy production responds to these needs through exchanges between territory and infrastructure depending on the opportunities for implementing equipment.

In more withdrawn industrial areas, such as Norman Street or the industrial areas in the north of Autoroute 20 in Dorval and Lachine, mainly wind turbines are suggested. Isolation of these structures in relation to the living environment would reduce nuisances.

In more sensitive places, especially those located near communities, light equipment energy producers are mainly suggested, including solar panels or equipment that uses piezoelectric energy.

On the highway itself, implementing a network of geothermal energy sensors when building the pillars of the new interchanges is suggested to service the neighboring districts.

Urban agriculture
Finally, although distinct from the industry, strictly speaking, agriculture is also perceived as a production chain whose implementation in the city entrance corridor could enhance the place's productive nature. Two implementation strategies are used. The first aims at completely restructuring the major industrial areas located near the Saint-Pierre interchange, Norman Street, and to the north of Autoroute 20 in Lachine. Here, the little valued industrial areas become agricultural plots whose boundaries come from historical territorial limits of the island of Montreal.

The second strategy aims to insert agriculture as a new production activity to the existing industrial areas. Here, greenhouse cultivation is often suggested. Greenhouses are implemented on rooftops of industrial or inside autonomous buildings.

Conclusion

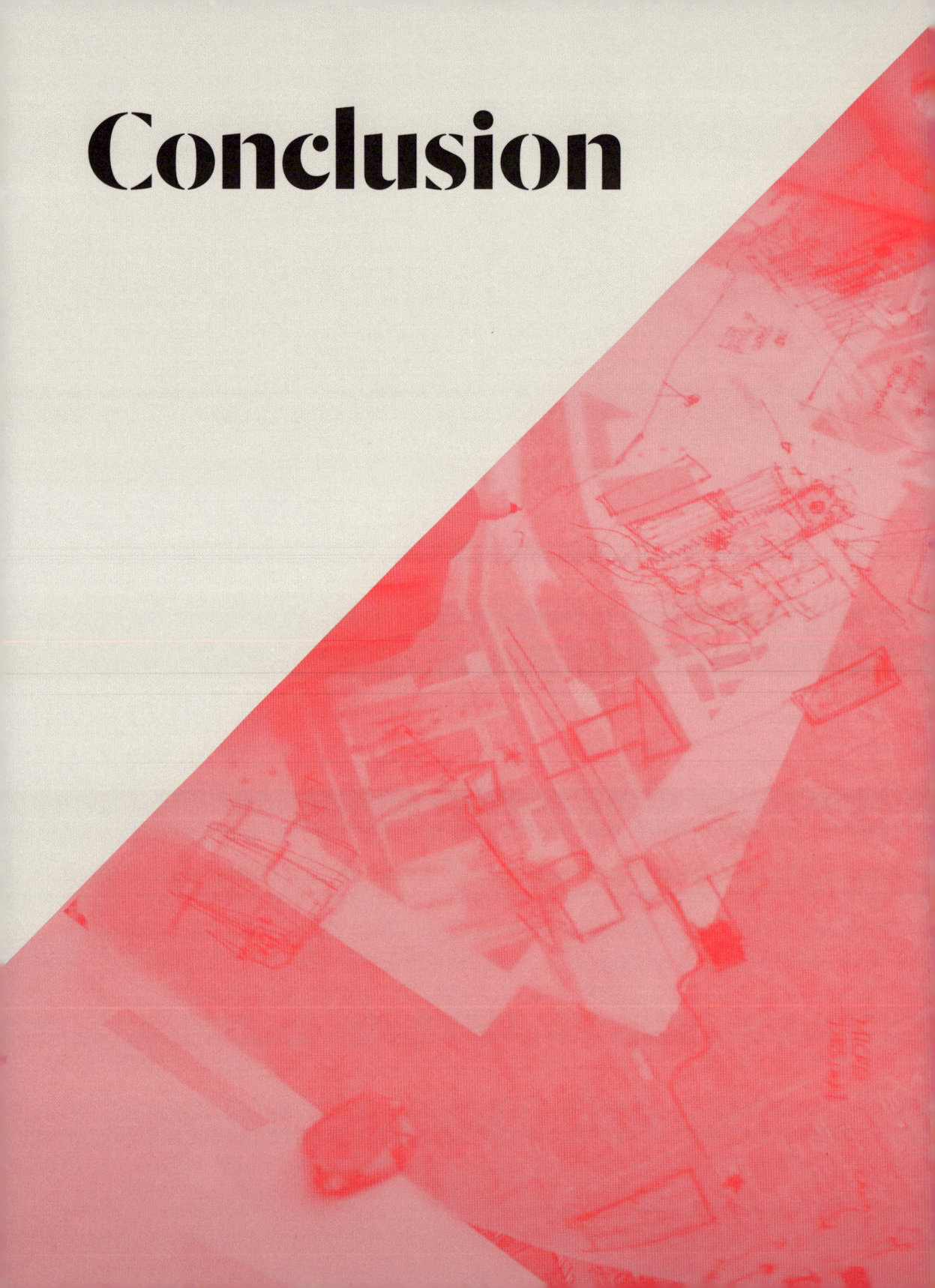

The YUL/MTL: Moving Landscapes project is based on the premise that the highway can have a positive landscape value in cities at the beginning of this century, as it partially was, over the last century, with the development in the United States of "parkways" and then "scenic roads". This observation has been effective in Europe since the 1990s[1]. On the North American side, this concern returned during the 2000s. The latter is expressed, especially through the notion of landscape urbanism[2] or, more recently, of infrastructural urbanism[3]. One common feature amongst the approaches is to posit that the highway, like other network infrastructures, is an area that offers opportunities to improve the quality of life, the urban environment, and enhance the areas around it.

It is important to emphasize that these positions have been built on a criticism of the dominant production model for infrastructures since 1950, in which transportation agencies are the only responsible ones for planning and building highway infrastructures, and where the user experience is summarized to criteria of fluidity, efficiency, and safety. It is worth noting, among other things, the influence of Jane Jacobs[4] in constructing criticism of this model through its opposition to various road projects in New York and then Toronto. Her efforts to counter urban highway projects largely contributed to the resurgence of interest for the city and more particularly for the public space quality of the central neighborhoods. This criticism was renewed in front of the observation that cities are becoming more and more splintered [5] and sprawled[6], and that the consumption of the territory generates residual areas[7] without specific use or non-places[8] special meaning. Faced with the constant and excessive progression of urban sprawl, and of the infrastructures that support it, the criticism has contributed doubt to the scope and relevance of urban infrastructures projects, and revealed

---

1. Lassus, B., op. cit.
2. Waldheim, C., (Ed), 2006. *The landscape urbanism reader*. New York: Princeton Architectural Press.
3. Hauck, T., R. Keller and V. Kleinekort (eds), 2011. *Infrastructural urbanism: addressing the in-between*. DOM publishers, Berlin.
4. Jacobs, J., 1961. *The death and life of great American cities*. Vintage Books, New York.
5. Graham, S. and S. Marvin, op. cit..
6. We think here about the concept of sprawled cities especially developed by Bernardo Secchi. See Secchi, B., 2009. *La ville du vingtième siècle*. Éditions Recherches, Paris.
7. Berger, A., 2006. Drosscape,in Waldheim C. (ed) *The landscape urbanism reader*. Princeton Architectural Press, New York, pp. 197-217.
8. Augé, M., 1992 *Non-lieux : introduction à une anthropologie de la surmodernité*. Seuil, Paris.

The YUL–MTL process raised awareness amongst numerous territorial stakeholders in terms of environmental, social and cultural aspects of the territory.

the highway's impacts on the natural, social, and built environment. In parallel, social awareness of these impacts was fueled in particular through the development of public collaboration mechanisms in the implementation of projects. Growing concerns over the past few decades relating to the environment and sustainable development contributed to affirm this social criticism.

The Autoroute 20 gateway corridor in Montreal is not immune to this context of urban crisis. It is the very example of a splintered territory caused by the transportation infrastructure. This entrance (and exit) pathway is today disconnected from the territory and represents a real barrier limiting both social and ecological exchanges. To this territorial division, administrative boundaries add a political complexity on the coherent planning of relations between infrastructures and urban areas.

Proposals coming from the *YUL/MTL: Moving Landscapes* project and their subsequent thoughts during the whole planning process are based on a reconceptualization of the role of the highway in the city. The highway infrastructure and urban contexts were called on, interpreted, and confronted in their complementarity. A complementarity that is primarily from an environmental viewpoint,[9] since highways form linear structures that may be assimilated, in certain circumstances, to ecological corridors – complementarity that is also functional. In fact, the interstitial spaces of the highway acquired a value of use either as an area of informal socialization – for example a skateboarding park – either as a park or public space as in the case of the Buffalo Bayou

---

9. Bélanger, P., 2010. Redefining Infrastructure, in Mostafavi M. and G. Doherty (eds) *Ecological Urbanism*. Lars Muller Publisher, Baden, pp. 332-349.

Facing uncertainties from issues associated with the territory transformation, ideation is an essential tool that ensures collective support for a project.

Promenade in Houston (United States) or of the Passeig de Garcia Faria in Barcelona (Spain).

Furthermore, the *YUL/MTL: Moving Landscapes* project contributed reflections on the role of the highway in the city by experimenting an approach that enables the development of a shared vision of the territory. Therefore, using landscape as a key concept of intervention, the YUL/MTL approach raised awareness among a large number of stakeholders for the importance of engaging infrastructure projects in a cross-sectional perspective that integrates, among other things, environmental, social, and cultural dimensions of the territory without mobilizing specific technical knowledge related to the management of road traffic and the construction of the civil engineering structures. To do this, the YUL/MTL project relies on two important considerations, those being:

- Collaboration through dialog between the local stakeholders on shared values and aspirations to develop a territorial vision giving a meaning to the territory ;
- Design ideation exercises (ideas competition and design workshop) as a source of illustration and enrichment of the territorial vision.

The international nature of the ideas competition and the design workshop reflected the global nature of the raised issues and allowed everyone to reflect on the possibilities of implementation, in the Montreal context, of planning concepts that are developing in various regions of the world. Although the benefits of this approach are still to be assessed in the long term, the project generated knowledge about the potential of using collaboration and ideation to become aware of the highway landscape issues and define an action plan.

In fact, *YUL/MTL: Moving Landscapes* enabled a real territory project by giving it a name, the *Autoroute 20 gateway corridor*, by drawing its geographic contours, and giving it planning goals that take meaning and shape across the images, themes, and actions they evoke. How do we maintain the existence of this territorial and landscape intention? How do we make it alive, evolving toward the implementation of concrete planning tools (ex.: landscape charter or master plan)?

The exploratory approach presented in this book has allowed for more demonstrations concerning the relevance of the ideation exercises in a planning process. It thus certifies the fact that an idea is not an end-purpose, but a tool. As a first step, this tool is allowed to illustrate the words and affirmations of a strategic position defined by the territorial stakeholders in the entrance of the metropolis. As a second step, the two ideation exercises (international ideas competition and WAT_UNESCO) have been central to enunciate the terms of urban planning for the implementation of specific projects. Lastly, they have scripted themes associated with the issues at hand, thus making a plurality of meaning for the territory. These three steps are the key components of an inclusive and collaborative approach that is essential to anchoring the *YUL/MTL: Moving Landscapes* project to a specific area.

Towards Implementation, A Few Inspiring Processes...

Over the last decade, several "soft" approaches emerged in Europe and North America to coordinate the planning process at a supra-local level in an incentivized manner rather than a regulatory and coercive way. Three examples can be pointed out as models to inspire the follow-up of the *YUL/MTL: Moving Landscapes* project: the International Building Exhibition (IBA) in Germany, the Design Pact of the Thames Gateway in southeastern England, and the strategic urban planning scheme of Antwerp (Belgium).

In the manner of a good practice catalyst, International Building Exhibitions (Internationale Bauaustellung - IBA) are held over a period of 10 years. Their objective is not to make an exhibition of what currently exists, but rather to facilitate the emergence of projects that are part of a drive for excellence. Therefore, it is the territory itself that becomes, at the end of the process, a living exhibition hall. Local stakeholders who commit to carry out projects under the IBA are guided by a project office (IBA-office). This office aims to inspire action through organizing public debates, international conferences, design workshop or exhibitions as well as offer technical planning expertise so that the interventions of local authorities reach the desired level of excellence and innovation. The project office also develops communication tools to offer a strong local, national, and international visibility to

Figure 5.1: Vague landscape impression, highway entrance from the Jacques-Cartier Bridge: Montreal (Canada) (Photo Credit: Philippe Poullaouec-Gonidec, CPEUM 2004)

the projects that are carried out. The IBA Emscher Park (1989-99)[10] and the IBA Hamburg (2006 – 13)[11] both contributed greatly to reviving their territory by producing a large number of projects of outstanding quality including the Gasometer in Oberhausen for the first and the IBA Dock for the second.

In the broad process of territorial revitalization of the Thames Gateway, the Design Pact[12] served as a commitment document for the main public stakeholders towards a high-quality urban environment production. Released in 2008, the principles statement included in this Pact is based both on design quality standards produced nationally and

---

10. Paquette, S., P. Poullaouec-Gonidec, M. Chenouda and D. Aubin, 2010. *IBA Emscher Park: Développement durable, culture et projets de territoire. Portrait de démarches québécoises et étrangères exemplaires*, document déposé au Ministère de la Culture, des Communications et de la Condition féminine, Chaire en paysage et environnement de l'Université de Montréal. (Online) www.agenda21c.gouv.qc.ca/ wp-content/uploads/2010/11/fiche-IBA-27oct2010.pdf (Page visited on May 5, 2014)
11. IBA Hamburg GmbH. IBA Hamburg Site, (Online). http://www.IBA-Hamburg.de/en/IBA-in-English.html (Page visited on May 5, 2014)
12. CABE, 2008. *The Thames Gateway design pact: making new things happen*. Collaboration draft. (Online) http://webarchive.nationalarchives.gov.uk/20110118095356/http://www.cabe.org.uk/ files/the-thames-gateway-design-pact.pdf (Page visited on May 5, 2014)

The challenge to elaborate a territory project for a city entrance lies in the capacity to produce significant meanings and landscape expressions to transcend the effects of a «generic city».

on a vision document[13] developed following a broad collaboration that highlighted the main identity elements and attractions of the region. Thus, the Design Pact, whose implementation continues until 2016, seeks the long-term enhancement and development of the local identity. The Center of Architecture and Built Environment (CABE, now integrated within the Design Council) was appointed to track this approach and assist local stakeholders in achieving quality objectives. It carries out this work through the service of its design review panel who participated in assessing projects. It should be noted that the design review panels are a tool that is increasingly used by cities to ensure project quality and their coherence within large regeneration development. In this respect, the experiences of Auckland, New Zealand and Toronto, Canada[14] should be noted. The design review panel set up by the latter within the Waterfront Toronto project targeting the redevelopment of its port area favored the emergence of high quality projects often chosen as a result of a design competition including the master plan of the waterfront central section designed by West 8 and a public space project named Sugar Beach from the landscape architect Claude Cormier.

    Finally, the strategic urban planning scheme of Antwerp, developed by Bernardo Secchi and Paula Vigano, is based on the production of seven images aimed at translating the city identity and channeling the direction of upcoming projects – to group them according to the raised

---

13. CABE, 2006. *New things happen. A guide to the future Thames Gateway*. (Online) http://webarchive.nationalarchives.gov.uk/20110118095356/http://www.cabe.org.uk/files/new-things-happen-report.pdf (Page visited on May 5, 2014)

14. Lemieux, D., 2008 *Imaginer, réaliser la ville du 21e siècle, Cahiers des bonnes pratiques en design. Cahier 1 : Neuf exemples internationaux pour inspirer le renouvellement de l'action publique en design urbain*. Ville de Montréal et ministère de la Culture, des Communications et de la Condition féminine du Québec : Montreal. (Online) http://mtlunescodesign.com/docs/projects/Cahier%20bonnes%20pratiques_01.PDF (Page visited on May 5, 2014)

stakes and thus giving a truly strategic scope to the actions of the city. These images in turn evoke the "city on water", the "eco-city", the "port-city", the "rail city", the "porous city", "villages and cities", and the "mega-city ". They reflect different aspects of town, but also different issues and areas of project opportunities. With this structure, the strategic planning scheme inspired the modification of the municipal structure to create specific teams to accompany the intervention on each of these themes.

These planning processes have several features in common. First, they revolve around a shared commitment over a long period (5 to 10 years) from multiple planning stakeholders towards issues or themes of intervention. Then, they seek complementarity between the orientations statement and the implementation of specific projects. Thus, orientations statements are open; they do not offer complete solutions or do not force action. Rather, they seek to develop innovation objectives to pursue the implementation of each project. Finally, they are based on approaches that accompany projects. Thus, local stakeholders remain responsible for the intervention on their territory, but they can ask advice and mentoring from a project office or specialized teams. It is also important to note that each step involves a sustained effort in terms of public communication. Since they primarily target an incentivized, stimulating reflection and inspiring action become core activities in implementing these projects.

Like these approaches of innovative planning, the *YUL/MTL: Moving Landscapes* project has the essence of a real territory project for Quebec's metropolis. It fits into a splintered territory in which several stakeholders are involved. It is based on a vision statement that describes both issues and identity elements that are common to them. It provides images of a possible future and themes that can stimulate dialogue and action. As pointed out by Bernardo Secchi and Paula Viganò, quoting historian Frances Yates, images "act on memory and guide the construction of the future"[15]. Too often, planning actions are limited to list priorities based on expert analyses of the economic, social, and environmental problems of the territory. Although it may result in a consensus on actions to be undertaken, this type of planning is most appropriate when the number of stakeholders is limited and the future risks are low. In plural societies and splintered cities facing uncertainty toward the future, both economically and environmentally, the construction of shared images becomes a mechanism and the unavoidable substrate to ensure the adhesion towards a project, for which collective value is essential to the vitality of living territories.

---

15. Secchi, B. and P. Vigano, 2011. A strategic scheme in Ariella Masboungi (ed) *Anvers, Faire aimer la ville*. Le Moniteur, Paris, pp 31-45.

Urban Landscape of Mobility

Beyond the singular process and means undertaken to address local issues, the *YUL/MTL: Moving Landscapes* project supports a new vision of the role and the meaning given to the highway and road transportation infrastructure fitting into urban environments increasingly concerned with value of the area. This vision is related to the landscape experience of the city.

At a time in which cities are looking for landscape recognition, it is important to recall, or even to recognize, that highway transportation infrastructures contributed to invent its landscapes[16], as did in Western cities the romantic writers of the 19th century[17] and the 20th century[18]. The peasant[19] and the "flâneur"[20] appear as characters who live and view the city as they wander the streets and use public transportation. The opposite of framed vistas deeply fitted into the representations (or figures) of the city[21], urban transportation brings stealth visions, fuzzy scrolling, foreground rupture, and background noise encompassing urban routes (Figure 5.1). Thus, the transported users experience of the city is kinetic and oscillates between visual acuity and the vague impression of a place that leads to musing.

It is possible to affirm that road and highway traffic in the city contributed to the invention of landscapes. Deployment of city horizons (e.g.: panoramas) along elevated highways, low-angle views, and continuous visual scrolling have accented the significance of urban landscapes expressions since the beginning of the 20th century. As a space of movement, the city offers multiple looks through which fixed and balanced visual framing takes precedence when reading sites. Framing is no longer the hegemony of appreciation. Mobility experience is quite different. It involves extending the visual field in all its dimensions and all its forms. From distance to proximity, from jamming to fixity, from slow to fast speed, the urban landscape of mobility reveals another face of the areas and their interpretation. Without wanting to

---

16. "Landscape is the result of a territory appreciation that develops based on the values (historical, aesthetic, ecological, economic, etc.), uses (residential, touristic, etc.) and sensory experiences (visual, auditory, olfactory, etc.). It concerns both a social and cultural enhancement phenomenon of an urban environment and the tangible and intangible expression of culture of individuals who live in it or use it" in Poullaouec-Gonidec, P. and S. Paquette, 2011. *Montréal en paysages*, Presses de l'Université de Montréal, Montreal.
17. Examples include Honoré de Balzac through his novel *Ferragus* (1833) and Charles Baudelaire in *Les fleurs du mal* (1861).
18. Several anthologies of poems recall the close link between literature and the city, for reference: *Montréal des écrivains* (1988), collective under direction of L. Dupré, B. Roy and F. Theoret, L'hexagone Edition and *Les poètes et la ville* (2006), collection of Jacques Reda, Edition Poesie/Gallimard.
19. Read about Aragon, 1926. *Le paysan de Paris*, Edition Gallimard.
20. Walter Benjamin (*Paris, capital du 19e siècle*, Edition de Cerf, Paris, 1989)
21. Read about Pousin, F., 2005. *Figures de la ville et construction du savoir*, Édition du CNRS-France, Paris.

praise mobility, the contemporary experience of movement significantly contributed to get the idea of landscape out of the formal constraint, i.e. mainly the visual framing. Thus, the landscape creates impressions and expressions of a place and/or of meaningful values for the community.

In highway requalification projects such as *YUL/MTL: Moving Landscapes*, the question of the landscape experience and its invention is central. How do we narrate the meaning given to the experience of crossing or entering and exiting the city? How do we stage the speed – the slow downs and stops – of movement to understand the traveled places? How do we articulate and structure the sequences through which the city unfolds? These questions necessarily refer to the contemporary understanding of urban landscapes. It must above all consider the urban experience in its entirety. Characterization of the urban landscape can no longer only be a directory of fixed frames on landmarks, boundaries, or urban nodes[22].

As previously mentioned, the highway travel experience is juxtaposed with others in a perspective of the infrastructure project that is open to the territory where the right-of-way is no longer a border, a barrier to the territory. Scenarios developed in the *YUL/MTL: Moving Landscapes* project are trying to provide a response through deploying evocative themes of collective aspirations and stakes facing both road users and local residents. These themes are a conjunction of sense, a common sense that allows uniting the highway infrastructure and its contexts. The contribution of singular sense is truly the key to landscape invention. Thus, the latter is the binder that participates in anchoring components of the territory (infrastructure, habitat, public space, etc.). The landscape thus becomes the unifying element for an urban mobility that is opened on the territory.

Under this perspective, the challenge of urban mobility with landscape value lies in the ability to produce an expressive, shareable, and collectively appropriable sense not only to plan a road or highway infrastructure, but a meaningful territory that will transcend the effects of a "generic city"[23].

---

22. These urban terms related to reading urban space refer to Lynch, K., 1960. *The image of the city*. M.I.T. Press, Cambridge.
23. Term borrowed from Architect Rem Koolhaas facing the observation of sprawled and splintered contemporary cities, see *Junkspace*, 2013. Edition Manuels Payot.

Bibliography

Ache, P., 2011. « Creating futures that would otherwise not be » - Reflections on the Greater Helsinki Vision process and the making of the metropolitans regions, *Progress in Planning*, 75: 155-192.

Allmendinger, P. and G. Haughton, 2009. Soft spaces, fuzzy boundaries, and metagovernance: the new spatial planning in the Thames Gateway, *Environment and Planning A*, 41: 617-633.

Appleyard, D., K. Lynch and J. Myer, 1964. *The view from the road*. M.I.T Press, Cambridge.

Aragon, 1926. *Le paysan de Paris*, Gallimard, Paris.

Augé, M., 1992. *Non-lieux : introduction à une anthropologie de la surmodernité*. Seuil, Paris.

Balzac, H., 1833. *Ferragus, chef des Dévorants*. Seuil, Paris.

Baudelaire, C., 1861. *Les fleurs du mal*, Gallimard, Paris.

Beck, H. and J. Cooper, 2000. *Denton, Corker, Marshall: rule playing and the ratbag element*. Birkhäuser, Basel.

Bélanger, P., 2010. Redefining Infrastructure, in Mostafavi M. and G. Doherty (Eds.) *Ecological Urbanism*. Lars Müller Publisher, Baden, pp. 332-349.

Benessaieh, K., Échangeur Turcot: un milliard gaspillé, selon Vision Montréal, *La Presse* (Online). (April 4, 2011) http://www.lapresse.ca/actualites/montreal/201104/04/01-4386340-echangeur-turcot-un-milliard-gaspille-selon-vision-montreal.php (Pages visited on May 5, 2014)

Benjamin, W., 1989. *Paris, capitale du XIXe*. Édition du Cerf, Paris.

Berger, A., 2006. Drosscape, in Waldheim C. (Ed.) *The landscape urbanism reader*. Princeton Architectural Press, New York, pp. 197-217.

Bishop, K. R., 1989. *Designing urban corridors*, Planning Advisory Service report no. 418, American Planning Association, Chicago.

Bisson, B., 2011. Le Comité de vigilance Turcot conseille d'élaguer le projet, *La Presse* (Online). (April 16) http://www.lapresse.ca/actualites/montreal/201104/16/01-4390502-le-comite-de-vigilance-turcot-conseille-delaguer-le-projet.php (Pages visited on May 5, 2014)

Bisson, B., 2010. La Ville propose un échangeur Turcot plus compact, *La Presse* (Online). (April 22) http://www.lapresse.ca/actualites/montreal/201004/22/01-4273008-la-ville-propose-un-echangeur-turcot-plus-compact.php (Pages visited on May 5, 2014)

Bourbeau, R., 1983. *Les accidents de la route au Québec 1926-1978. Étude démographique et épidémiologique*, Presses de l'Université de Montréal, Montréal.

Bureau d'audiences publiques sur l'environnement, 2009. *Projet de reconstruction du complexe Turcot à Montréal, Montréal-Ouest et Westmount. Rapport d'enquête et d'audience publique*. (Online) http://www.bape.gouv.qc.ca/sections/rapports/publications/bape262.pdf (Pages visited on May 5, 2014)

CABE, 2008. *The Thames Gateway design pact: making new things happen*. Consultation draft. (Online) http://webarchive.nationalarchives.gov.uk/20110118095356/http://www.cabe.org.uk/files/the-thames-gateway-design-pact.pdf (Pages visited on May 5, 2014)

CABE, 2006. *New things happen. A guide to the future Thames Gateway*. (Online) http://webarchive.nationalarchives.gov.uk/20110118095356/http://www.cabe.org.uk/files/new-things-happen-report.pdf (Pages visited on May 5, 2014)

Communauté métropolitaine de Montréal, 2011. *Un Grand Montréal attractif, compétitif et durable. Plan métropolitain d'aménagement et de développement* (Online) http://pmad.ca/fileadmin/user_upload/pmad2011/documentation/20111208_pmad.pdf (Pages visited on May 5, 2014)

De Jonge, J., 2009. *Landscape architecture between politics and science. An integrative perspective on landscape planning and design in the network society*. PhD thesis Wageningen University.

Dupré, L., B. Roy and F. Théoret (Eds.), 1988. *Montréal des écrivains*, Édition L'Hexagone, Montréal.

Gariépy, M., P. Lewis., N. Valois and L. Desjardins, 2006. *Le cadrage paysager des entrées routières de Montréal*. Ministère des Transports du Québec, Québec.

Gourdon, J.-L., C. Werquin and A. Demangeon., 2000. *Boulevards, rondas, parkways... des concepts de voies urbaines*. Ministère de l'équipement, du logement, des transports et du tourisme. Centre d'études sur les réseaux, les transports, l'urbanisme et les constructions publiques (CERTU), Lyon.

Graham, S. and S. Marvin, 2001. *Splintering Urbanism, Networked Infrasstructures, technological mobilities and the urban condition*. Routledge, Londres.

Guthrie, J., 2010. Pour un projet Turcot réduit et plus humain, *Journal Métro* (Online). (March 25) http://journalmetro.com/actualites/montreal/33154/pour-un-projet-turcot-reduit-et-plus-humain/ (Pages visited on May 5, 2014)

Hanna, D. B., 1993. *Transport des personnes et développement du territoire de l'agglomération montréalaise : Un essai d'interprétation historique*. Prepared for Service de la planification du territoire de la Communauté urbaine de Montréal.

Hauck, T., R. Keller and V. Kleinekort (Eds.), 2011. *Infrastructural urbanism: addressing the in-between*. DOM publishers, Berlin.

Hung, Y.-Y. and SWA Group, 2011. *Landscape infrastructure. Case studies by SWA*, Birkhäuser, Basel.

IBA Hamburg GmbH. BA Hamburg, (Online). http://www.iba-hamburg.de/en/iba-in-english.html (Page visited on May 5, 2014)

Jacobs, J., 1961. *The death and life of great American cities*. Vintage Books, New York.

Jacobs, P., P. Poullaouec-Gonidec, B. St-Denis, C. Bélanger, D. Hadj-Hamou and L. Lévesque, 1998. *Étude de caractérisation et de requalification des paysages d'entrée de la capitale du Québec : Le corridor Duplessis*. Rapport final. Chair in Landscape and Environmantal Design at Université de Montréal, Montréal.

Lassus, B., 2014, Évocations de la longue histoire de l'autoroute Européenne E6.

Lassus, B., 1998. *The Landscape Approach*. University of Pennsylvania Press, Philadelphia.

Legault, G. R., 2002. *La ville qu'on a bâtie, trente ans au service de l'urbanisme et de l'habitation à Montréal, 1956-1986*. Liber, Montréal.

Lemieux, D. 2008. *Imaginer, réaliser la ville du 21ᵉ siècle, Cahiers des bonnes pratiques en design. Cahier 1 : Neuf exemples internationaux pour inspirer le renouvellement de l'action publique en design urbain*. Ville de Montréal et ministère de la Culture, des Communications et de la Condition féminine du Québec : Montréal. (Online) http://mtlunescodesign.com/docs/projects/Cahier%20 bonnes%20pratiques_01.pdf (Page visited on May 5, 2014)

Lortie, A. (Ed.), 2004. *Les années 1960, Montréal voit grand*. Centre Canadien d'Architecture, Montréal.

Lynch, K., 1960. *The image of the city*. M.I.T. Press, Cambridge.

Koolhaas, R., 2011. *Junkspace: repenser radicalement l'espace urbain*. Payot, Paris.

Michaelson, J., G, Toth, and R. Espiau, 2008. *Great Corridors, Great Communities: The Quiet Revolution in Transportation Planning*. Project for Public Spaces, New-York.

Ministère des Transports du Québec. Capsules historiques (Online) https://www.mtq.gouv.qc.ca/portal/page/portal/100ans/capsules_historiques (Page visited on May 5, 2014)

Ministère des transports du Québec. Répertoire des autoroutes du Québec (Online) http://www1.mtq.gouv.qc.ca/fr/repertoire_autoroute/autoroute.asp (Page visited on November 6, 2013)

Niordson, H. and T. Erlandson, 2014. in Lassus, B (Ed.) Évocations de la longue histoire de l'autoroute Européenne E6.

Noppen, L., 2001. *Du chemin du Roy à la rue Notre-Dame, mémoires et destins d'un axe est-ouest à Montréal*. Ministère des Transports du Québec, Montréal.

Paquette, S., P. Poullaouec-Gonidec, M. Chenouda and D. Aubin, 2010. *IBA Emscher Park: Développement durable, culture et projets de territoire. Portrait de démarches québécoises et étrangères exemplaires*, document submitted to Ministère de la Culture, des Communications et de la Condition féminine, Chaire en paysage et environnement de l'Université de Montréal. (Online) www.agenda21c.gouv.qc.ca/wp-content/uploads/2010/11/fiche-IBA-27oct2010.pdf (Page consulted on May 5, 2014)

Paquette, S., P. Poullaouec-Gonidec and G. Domon, 2009. *Québec Landscape Management Guide : Reading, Understanding, and Enhancing the Landscape*, Ministère de la Culture, des Communications et de la Condition féminine, Québec.

Parcs Canada, 2004. Lieu historique national du Canada du Canal-de-Lachine Plan directeur (Online) http://www.pc.gc.ca/fra/lhn-nhs/qc/canallachine/docs/plan1.aspx (Page visited on May 5, 2014)

Poullaouec-Gonidec, P. and S. Paquette, 2011. *Montréal en paysages*, Presses de l'Université de Montréal, Montréal.

Pousin, F., 2005. *Figures de la ville et construction du savoir*, Édition du CNRS-France, Paris.

Radde, B., 1993. *The Merritt Parkway*. Yale University Press, New Haven.

Rauws, W and T. van Dijk, 2013. A design approach to forge visions that amplify paths of periurban development, *Environment and Planning B: Planning and Design*, 40 : 254-270.

Réda, J. (Éd.), 2006. Les poètes et la ville, Édition Poèsie / Gallimard.

Secchi, B. and P. Vigano, 2011. Un schéma stratégique, in Masboungi, A. (Éd.) *Anvers, Faire aimer la ville*, Le Moniteur, Paris, pp 31-45.

Secchi, B., 2009. *La ville du vingtième siècle*. Éditions Recherches, Paris.

Société de l'assurance automobile du Québec (SAAQ), 2012. Bilan routier 2012. (Online) http://www.saaq.gouv.qc.ca/publications/prevention/bilan_routier_2012/bilan_routier.pdf (Page visited on November 4, 2013)

St-Denis, B., C. Marcoux, and M.-C. Paradis, 2003. *Cadrage des entrées à la capitale nationale de Québec*. Report submitted to ministère des Transports du Québec. Chaire en paysage et environnement de l'Université de Montréal, Montréal.

United States Department of Transportation, Federal Highway Administration, 2007. *What is CSS?* (Online) www.fhwa.dot.gov/context/ (Page visited on April 29, 2014)

Van Dijk, T., 2011. Imagining future places: How designs co-constitute what is, and thus influence what will be, *Planning Theory*, 10 (2): 124-143.

Vigano, P., 2012. *Les territoires de l'urbanisme, le projet comme producteur de connaissance*. MétisPresses, Genève.

Ville de Montréal. Le parcours riverain: Chemin de Lachine et du bord du lac Saint-Louis (Online) http://ville.montreal.qc.ca/portal/page?_pageid=8817,99661605&_dad=portal&_schema=PORTAL (Page visited on May 5, 2014).

Ville de Montréal, Service d'urbanisme, 1948. *Une Autostrade Est-Ouest*. Montréal, Ville de Montréal.

Waldheim, C., (Ed.) 2006. *The landscape urbanism reader*. Princeton Architectural Press, New York.

Mapping resources

Bouchette, J., 1831. To his most Excellent Majesty, King William IV. This topographical map of the district of Montreal, Lower Canada (cartographic document). 1:175,000, Bibliothèque et Archives nationales Québec. http://services.banq.qc.ca/sdx/cep/document.xsp?id=0000090116 (Page visited on July 10, 2013)

Département de la Défense nationale, 1967. Carte topographique du Canada, 31-H-05-g, Lachine (cartographic document). 1:25,000, Bibliothèque et Archives nationales Québec. http://services.banq.qc.ca/sdx/cep/document.xsp?id=0002671938 (Page visited on July 12, 2013)

Département de la Défense, 1915. Carte topographique du Canada, 31-H-05, Lachine (cartographic document). 1:63,360, Bibliothèque et Archives nationales Québec. http://services.banq.qc.ca/sdx/cep/document.xsp?id=0002684422 (Page visited on July 12, 2013)

Département des Mines et des Ressources, 1940. Carte topographique du Canada, 31-H, Montréal (cartographic document). 1:253,440, Bibliothèque et Archives nationales Québec. http://services.banq.qc.ca/sdx/cep/document.xsp?id=0002670072 (Page visited on July 12, 2013)

Ministère des Ressources naturelles, 1998. Carte topographique du Québec, 31-H-05-200-0202 (cartographic document). 1:20,000, Bibliothèque et Archives nationales Québec. http://services.banq.qc.ca/sdx/cep/document.xsp?id=0002683860 (Page visited on July 12, 2013)

Sitwell, H.S., 1869. Contoured plan of Montreal and its environs, Quebec (cartographic document). 1:2,500, Bibliothèque et Archives nationales Québec. http://services.banq.qc.ca/sdx/cep/document.xsp?id=0000321499 (Page visited on July 10, 2013)

Authors

Philippe Poullaouec-Gonidec is the Director of the Chair in landscape and environmental design as well as the UNESCO Chair in landscape and environmental design of the University of Montreal. Landscape Architect and artist of the environment, he is a professor at the School of landscape architecture of the University of Montreal, where he was Director between 1991 and 1996. He manages a scientific network of more than two dozen academic institutions around the world to promote interdisciplinary research and an intercultural dialog on the development of urban landscapes. In 2005, he received one of the five prestigious Trudeau research awards (winner of the Trudeau Foundation) in recognition of his outstanding contribution to questions of public interest in the field of landscape architecture. Pursuing his interest in contemporary art, he was co-founder of the International Garden Festival of Metis. In 2007, he was named Chevalier de l'Ordre des Arts et des Lettres in France.

Sylvain Paquette has been a researcher at the Chair in landscape and environmental design as well as at the UNESCO Chair in landscape and environmental design of the University of Montreal since 2005. Graduated with a Ph. D. (planning, 2001) from the University of Montreal, he is currently an associate professor in the School of landscape architecture of this institution. His works are within the field of landscape sociology and address in this sense the question of landscape as a phenomenon of social and cultural enhancement for the urban and suburban territories. These contributions to research spread at the national and international level participated in renewing approaches in landscape studies and territory development, as far as conceptual frameworks, methodological perspectives, landscape strategies and management tools.

Patrick Marmen is a research officer at the Chair in landscape and environmental design of the University of Montreal and studio instructor at the School of landscape architecture of this institution. Interested in the development and improvement of the planning process that promote the quality of design in architecture and urban design, he primarily acted as a professional counselor for the development and coordination of the YUL/MTL international competition of ideas: Moving Landscapes. He holds a master of architecture (M. Arch) from the University of Laval.

The **Chair in landscape and environmental design of the University of Montreal** is a unique research laboratory in Canada that has been consistently committed for almost 20 years to a contextualized research in the area of landscape and urban planning through the construction of plural knowledge that appeals to urban planning disciplines, human and natural sciences. In partnership with public agencies, it conducts a reflection that aims to surpass traditional approaches of landscape analysis and intervention. The agency is first and foremost a sustained research-action structure and projects experimentation which purpose is the development of knowledge and means of intervention in the areas of landscape.

The **UNESCO Chair in landscape and environmental design of the University of Montreal** is a scientific agency that deploys research and higher education on the development of cities and their landscapes. To address various issues of preservation, enhancement and development of urban landscapes in different regions of the world, it counts on the realization of Workshops_atelier/terrain (WAT_UNESCO). Thus, founded on a network of international cooperation between more than 20 academic institutions in six regions of the world (Europe, North Africa, Middle East, Asia, South America, and North America), it helps cities to solve urban development problems (redefining neighborhoods, industrial and periurban areas) through realizing urban planning visions in consultation with the local communities.

Additional credits

p. 88: Brown & Storey, Canada
p. 97: Brown & Storey, Canada
p. 124: Catalyse urbaine, Canada
p. 145: WAT_UNESCO Montreal – Team 1A
p. 168: Catalyse urbaine, Canada
p. 173: Johnathan Yue, Singapore
p. 179: Ghazal Jafari + Ali Fard, Canada
p. 192: dlandstudio, United States
p. 211: David Garcia Studio, Denmark